EIGHT LECTURES ON THEORETICAL PHYSICS

MAX PLANCK

TRANSLATED BY

A. P. WILLS

WITH A NEW INTRODUCTION AND NOTES

BY

PETER PESIC

ST. JOHN'S COLLEGE

SANTA FE, NEW MEXICO

DOVER PUBLICATIONS, INC.

Mineola, New York

The editor wishes to thank Stephen Brush, Gerald Holton, Allan Needell, and Curtis Wilson for their helpful advice.

Copyright

Published in Canada by General Publishing Company, Ltd., 30 Lesmill Road, Don Mills, Toronto, Ontario.
Published in the United Kingdom by Constable and Company, Ltd., 3 The Lanchesters, 162–164 Fulham Palace Road, London W6 9ER.

Bibliographical Note

This Dover edition, first published in 1998, is an unabridged republication of *Eight Lectures on Theoretical Physics, Delivered at Columbia University in 1909*, first published by The Columbia University Press in 1915, as Publication Number 3 of the *Ernest Kempton Adams Fund For Physical Research*. A new introduction and notes have been prepared for this edition by Peter Pesic.

Library of Congress Cataloging-in-Publication Data

Planck, Max, 1858–1947.
[Acht vorlesungen über theoretische physik. English]
Eight lectures on theoretical physics / Max Planck ; translated by A.P. Wills ; with a new introduction and notes by Peter Pesic.
p. cm.
Originally published: New York : Columbia University Press, 1915.
Includes bibliographical references.
ISBN 0-486-69730-4 (pbk.)
1. Physics. I. Wills, A. P. II. Title.
QC71.P613 1998
530—dc21 98-10716
CIP

Manufactured in the United States of America
Dover Publications, Inc., 31 East 2nd Street, Mineola, N.Y. 11501

PREFACE TO ORIGINAL EDITION.

The present book has for its object the presentation of the lectures which I delivered as foreign lecturer at Columbia University in the spring of the present year under the title : "The Present System of Theoretical Physics." The points of view which influenced me in the selection and treatment of the material are given at the beginning of the first lecture. Essentially, they represent the extension of a theoretical physical scheme, the fundamental elements of which I developed in an address at Leyden entitled: "The Unity of the Physical Concept of the Universe." Therefore I regard it as advantageous to consider again some of the topics of that lecture. The presentation will not and can not, of course, claim to cover exhaustively in all directions the principles of theoretical physics.

THE AUTHOR.

BERLIN, 1909.

TRANSLATOR'S PREFACE.

At the request of the Adams Fund Advisory Committee, and with the consent of the author, the following translation of Professor Planck's Columbia Lectures was undertaken. It is hoped that the translation will be of service to many of those interested in the development of theoretical physics who, in spite of the inevitable loss, prefer a translated text in English to an original text in German. Since the time of the publication of the original text, some of the subjects treated, particularly that of heat radiation, have received much attention, with the result that some of the points of view taken at that time have undergone considerable modifications. The author considers it desirable, however, to have the translation conform to the original text, since the nature and extent of these modifications can best be appreciated by reference to the recent literature relating to the matters in question.

A. P. WILLS.

CONTENTS.

INTRODUCTION

Max Planck's long life (1858–1947) encompassed the heights of scientific accomplishment, two cataclysmic wars, and the tragic deaths of his four children. Near its end he wrote: "Our every starting-point must necessarily be something relative. . . . Our task is to find in all these factors and data, the absolute, the universally valid, the invariant, that is hidden in them."[1] The search for the absolute was the guiding theme of Planck's life. It began with "a revelation, the first law I knew to possess absolute, universal validity, independently from all human agency: the principle of the conservation of energy." But most of all he was drawn to the second law of thermodynamics: "there are processes in nature which in no possible way can be made completely reversible."[2] In the concept of *irreversibility* Planck expressed his vision of the absolute which led him to discover quantum theory in 1900.[3]

Planck's path to the quantum began with the assumption that the irreversible increase of entropy was as absolute a law as the conservation of energy. Beginning in 1897 he turned to the problem of the spectrum of electromagnetic radiation in a "black

[1]Max Planck, *Scientific Autobiography and Other Papers* (New York: Philosophical Library, 1949), p. 47; original text in Max Planck, *Physikalische Abhandlungen und Vorträge* (Braunschweig: Friedrich Vieweg & Sohn, 1958), hereafter cited as *PAV*, noting volume and page numbers, in this case 3.374-401.

[2]In 1897 Planck first published his *Treatise on Thermodynamics*, tr. Alexander Ogg (Dover, 1945) [translation of seventh German edition, 1922].

[3]See the extremely helpful introduction by Allan A. Needell to Max Planck, *The Theory of Heat Radiation* (New York: Tomash/American Institute of Physics, 1988), pp. xi–xlv as well as his "Irreversibility and the Failure of Classical Dynamics: Max Planck's Work on the Quantum Theory 1900–1915" (unpublished doctoral dissertation, Yale University, 1980).

body," a box with perfectly absorbing walls heated to incandescence, basically an oven with a small hole which allows some of the radiation to escape.[4] The light from an ordinary light bulb gives a fair approximation to the radiation emerging from an idealized black body. Kirchhoff had argued in 1859 that such radiation must have certain universal, absolute qualities regardless of the material of the box. Planck decided to use an idealized model to determine these absolutes. In his thought experiment he imagines inside the box a single "resonator," a pair of equal and opposite charges connected by a hypothetical spring. This resonator comes to equilibrium immersed in the field of electrodynamic radiation; Planck applied Maxwell's equations to find the distribution of energy in the box.

Sharp criticism by Ludwig Boltzmann caused him to change directions. Boltzmann objected that Planck's model used only perfectly *reversible* fundamental processes and hence could never explain the *irreversible* transformation of energy into radiant heat in the glowing oven. Instead, the molecular chaos of the initial state leads to irreversibility. Planck saw the force of these arguments and, in 1898, began to apply to radiation what he had learned from Boltzmann. However, Planck still held that entropy could never decrease, even though Boltzmann thought it was possible, though unlikely.

By October 1900 Planck's colleagues Rubens and Kurlbaum presented important new data on the spectrum of radiation emerging from the black body. In order to fit that data Planck was forced to modify Boltzmann's techniques by requiring that

[4]For general background see Max Jammer, *The Conceptual Development of Quantum Mechanics* (New York: McGraw-Hill, 1966), pp. 1–28; Armin Hermann, *The Genesis of Quantum Theory (1899–1913)* (Cambridge: MIT Press, 1971); Hans Kangro, *Early History of Planck's Radiation Law* (London: Taylor & Francis, 1976); Jagish Mehra and Helmut Rechenberg, *The Historical Development of Quantum Theory* (New York: Springer-Verlag, 1982), vol. 1, pt. 1, pp. 23–99, and pt. 2, pp. 613–639; Christa Jungnickel and Russell McCormmach, *Intellectual Mastery of Nature* (Chicago: University of Chicago Press, 1986), vol. 2, pp. 248–252, 260–268; Thomas S. Kuhn, *Black-Body Theory and the Quantum Discontinuity 1894–1912* (Chicago: University of Chicago Press, 1987), pp. 3–71.

the energy of the resonator "be composed of a well-defined number of equal parts" or "energy elements", rather than "a continuously divisible quantity," as classical electromagnetic theory implied.[5] This was the "most essential point" of the theory he announced on December 14, 1900, which fit the data of Planck's experimental colleagues very well and also predicted the value of several important atomic constants.[6] The full magnitude of his discovery went relatively unnoticed for several years, during which Planck published a fuller account of his application of thermodynamics to radiation.[7]

By 1907 a number of physicists, including H. A. Lorentz, realized the significance of Planck's work. Albert Einstein used Planck's quantum postulate in an even more radical way in his light-quantum paper of 1905. In 1908 Lorentz invited Planck to Leiden to lecture on "The Unity of the Physical World Picture."[8] In this lecture Planck began to give a deeper account of his quantum theory based on his assertion that in order to reveal the

[5]The writings of Martin J. Klein emphasize Planck's use of thermodynamics: "Max Planck and the Beginnings of the Quantum Theory," *Archive for History of Exact Sciences* **1**, 459–479 (1962); "Planck, Entropy, and Quanta, 1900–1906," *The Natural Philosopher* **1**, 83–108 (1963); "Thermodynamics and quanta in Planck's work," *Physics Today* **19**:11, 23–32 (1966); "The Beginnings of the Quantum Theory" in *History of Twentieth Century Physics: Proceedings of the International School of Physics "Enrico Fermi," Course LVII* (New York: Academic Press, 1977), pp. 1–39. See also the summary given in Klein's biographical study *Paul Ehrenfest: Volume 1, The Making of a Theoretical Physicist* (New York: American Elsevier, 1970), pp. 217–230.

[6]This paper, "On the theory of the Energy Distribution Law of the Normal Spectrum" (*PAV* 1.698–706), along with the brief announcement Planck gave on October 19, 1900 (*PAV 1.687–689*), is included in *Planck's Original Papers in Quantum Physics*, ed. Hans Kangro, tr. D. ter Haar and Stephen G. Brush (London: Taylor & Francis, 1972), pp. 37–45; the phrases cited are on p. 40.

[7]In 1906 Planck gave a detailed treatment of the black body in his *The Theory of Heat Radiation*, tr. Morton Masius (Dover, 1993) [translation of second German edition, 1913]. For early responses see Elizabeth Garber, "Some Reactions to Planck's Law, 1900–1914," *Studies in History and Philosophy of Science* **7**, 89–126 (1976) and Kuhn, *Black-Body Theory*, pp. 134–140.

[8]Reprinted under the title "The Unity of the Physical Universe" in Max Planck, *A Survey of Physical Theory* (Dover, 1991), pp. 1–26; *PAV* 3.6-29.

absolute any trace of anthropomorphism had to be removed from physical theory.

In 1909 Planck delivered a series of eight lectures at Columbia University giving an overview of the new situation in physics, in which he had played such a signal part, and carrying further the ideas he had broached at Leiden. Lorentz had already visited Columbia to lecture in 1906; at the time Planck came his ideas were just beginning to enter the curriculum of American universities, particularly at the University of Chicago under Robert A. Millikan's aegis.[9] Planck's Columbia lectures began on April 23 and were given over four consecutive weekends. Each Friday a general lecture preceded a more technical lecture the following day. These lectures, which he repeated in Leipzig in 1910, give a fascinating perspective on how he understood the new departures which he had initiated in 1900.

The first, third, fifth, and sixth lectures present his account of these developments. The reader is given an invaluable opportunity to witness Planck's thought processes on the level of philosophical principles as well as in their application to the physical processes on the microscopic and macroscopic scales. Throughout, he calls attention to anything "anthropomorphic" and moves toward invariant, universal principles. He shows the centrality of statistical ideas in purifying thermodynamics of the limitations of ordinary practical applications. From this, he points to the distinction between reversible and irreversible processes as the touchstone of the new physics. For Planck, the unavoidability of irreversible processes leads toward the introduction of atomicity at the fundamental level of physics. Nowhere else in his writings

[9]For detailed discussion of the American response to quantum theory at that time see Katharine Russell Sopka, *Quantum Physics in America: The years through 1935* (New York: Tomash/American Institute of Physics, 1988), pp. 26–46. For the influence of Millikan see Gerald Holton, "On the Hesitant Rise of Quantum Physics Research in the United States," in *Thematic Origins of Scientific Thought: Kepler to Einstein*, revised edition (Cambridge: Harvard University Press, 1988), pp. 147–187. Planck's 1909 lectures were translated into English by Albert P. Wills (1873–1937), a professor of mathematical physics at Columbia University who had studied at Göttingen and Berlin in 1898–1899.

does Planck present his argument in a more trenchant or compelling form.

In the second and fourth lectures Planck shows how these new ideas of statistical mechanics have transformed the understanding of the chemical physics of dilute solutions and of monatomic gases. Planck especially acknowledges the seminal work of Josiah Willard Gibbs not only as preeminent in America but also as one of "the most famous theoretical physicists of all times." The seventh lecture turns to the principle of least action, which Planck considers as essential to all reversible processes. His reflections concern the relation of these processes to general irreversible processes and show how he attempts to reconcile them. The final lecture is an account of the theory of special relativity, which Planck had early championed and whose creator was still working at the Swiss Patent Office.[10] Soon after Einstein had published his seminal paper (1905) Planck already discerned its profound importance as a revolution in thought comparable only to that brought about by the Copernican system.[11] He found in relativity theory not relativism but instead the highest expression of the invariant, the kind of eternal law that physics can attain when it purges itself of what is merely human and limited.

Planck's reflective prose clarifies the philosophical issues underlying his presentation. Though its overarching laws are absolute, nature's microscopic disorder implies that what happens

[10]Stanley Goldberg, "Max Planck's Philosophy of Nature and His Elaboration of the Special Theory of Relativity," *Historical Studies in the Physical Sciences* **7**, 125–160 (1976). Einstein resigned from the Patent Office on July 6, 1909, shortly after Planck's lectures ended.

[11]Einstein's 1905 paper is reprinted in *The Principle of Relativity* (Dover, 1952), pp. 37–65.

[12]In contrast, Thomas S. Kuhn maintained that Planck's initial approach was essentially classical, and that Einstein really was the originator of the radical departures of quantum theory; see his *Black-Body Theory*. For responses to Kuhn's interpretation see Martin J. Klein, Abner Shimony, and Trevor J. Pinch, "Paradigm Lost? A Review Symposium" *Isis* **70**, 429–440 (1979); Needell, "Irreversibility" and "Introduction"; and Peter Galison, "Kuhn and the Quantum Controversy," *British Journal for the*

today can never be undone. Written during a period when Planck was reconsidering his work, the lectures also show his critical attitude towards classical physics.[12] Though he hated to introduce fundamental discontinuities[13] Planck did count the states of the radiation inside the black body in a radically new way. Although he used Boltzmann's statistical techniques he insisted on an "elementary disorder" which prevents any possible decrease of entropy. Holding to this absolute sense of the second law and to the impossibility of eliminating this disorder Planck went beyond Boltzmann by treating the modes of radiation as wholly indistinguishable.

Planck's counting exemplifies a new concept of *identicality*, which joins *equality* of observable physical quantities (like mass or charge) to a radical *indistinguishability* that can confuse space-time histories.[14] This is a profound theme of the quantum theory that Ladislas Natanson and Paul Ehrenfest were among the first to notice (1911).[15] Though Einstein had been bold in grasping the implications of the quantum hypothesis, he remained conservative in wishing "to follow the course of individual atoms and forecast their activities," at least in principle.[16] Ironically, Planck's devotion to the absolute did not make him join Einstein in requir-

Philosophy of Science **22**, 71–85 (1981). According to Eugene Wigner, "that Planck believed in the quantum jump is evident" yet Planck did not believe "in the details in any derivation of [his equation], and this was natural since the physics of that time was full of contradictions," as in the case of the Bohr atom; cited in *Some Strangeness in the Proportion*, ed. Harry Woolf (Reading, MA: Addison-Wesley, 1980), p. 194. Kuhn replied to his critics in "Revisiting Planck," *Historical Studies in the Physical Sciences* **14**, 231–252 (1984), reprinted as the afterword to his *Black-Body Theory*, pp. 349–370.

[13]As Planck wrote to Paul Ehrenfest in 1915, "I hate discontinuity of energy even more than discontinuity of emission"; cited in Kuhn, *Black-Body Theory*, p. 253.

[14]See P. Pesic, "The Principle of Identicality and the Foundations of Quantum Theory. I. The Gibbs Paradox; II. The Role of Identicality in the Formation of Quantum Theory," *American Journal of Physics* **59**, 971–974, 975–979 (1991).

[15]See Abraham Pais, *Inward Bound* (Oxford: Oxford University Press, 1986), p. 283 and Martin J. Klein, "Ehrenfest's contributions to the development of quantum statistics," *Proceedings of the Amsterdam Academy* **B62**, 41–62 (1959).

ing this fundamental determinism. Planck looked to certain fundamental laws, especially of thermodynamics, more than to mechanically distinguishable atoms with individual histories. At the behest of those laws Planck was ready to suspend the identity of individual atoms, even though he was entering a new realm with whose strangeness he would long struggle. After all, from his youth he had surmised that the realm of the absolute contained more than classical atomic physics dreamt of.[17]

Planck's lectures are imbued with his intense integrity and thoughtfulness. This cautious, conservative man made the first steps towards a new view of the world. Altogether one feels in the presence of a great and admirable mind. "The state of mind which enables a man to do work of this kind is akin to that of the religious worshipper or the lover; the daily effort comes from no deliberate intention or program, but straight from the heart."[18] Thus Einstein spoke about his friend, praising his ardent quest for the "pre-established harmony" hidden behind the things of this world. These lectures allow us to hear Planck describe what he thought he was doing and how he viewed the whole scene of physics in that first amazing decade of the twentieth century.

Peter Pesic

[16]Quoted in Max Planck, *Where is Science Going?* (Woodbridge, CT: Ox Bow Press, 1981), p. 202.

[17]From his student days Planck had considered atomic theory critically and entertained the suspicion that matter was really continuous; see *PAV* 1.163 (1881), 3.2 (1894), J. L. Heilbron, *The Dilemmas of an Upright Man: Max Planck as Spokesman for German Science* (Berkeley: University of California Press, 1986), pp. 9–17, and Kuhn, *Black-body Theory*, pp. 23–26. For Planck's youthful studies of philosophy see Jost Lemmerich, "Der Student Max Planck hörte Philosophie bei Karl Prantl," *Bibliothek und Wissenschaft* **25**, 284–300 (1991).

[18]Albert Einstein, "Principles of Research" [address delivered on Planck's 60th birthday, 1918] in his *Ideas and Opinions* (New York: Crown Publishing Company, 1982), pp. 224–227; see Gerald Holton's insightful commentary in *Thematic Origins*, pp. 371–398.

FIRST LECTURE.

Introduction: Reversibility and Irreversibility.

Colleagues, ladies and gentlemen: The cordial invitation, which the President of Columbia University extended to me to deliver at this prominent center of American science some lectures in the domain of theoretical physics, has inspired in me a sense of the high honor and distinction thus conferred upon me and, in no less degree, a consciousness of the special obligations which, through its acceptance, would be imposed upon me. If I am to count upon meeting in some measure your just expectations, I can succeed only through directing your attention to the branches of my science with which I myself have been specially and deeply concerned, thus exposing myself to the danger that my report in certain respects shall thereby have somewhat too subjective a coloring.

From those points of view which appear to me the most striking, it is my desire to depict for you in these lectures the present status of the system of theoretical physics. I do not say: the present status of theoretical physics; for to cover this far broader subject, even approximately, the number of lecture hours at my disposal would by no means suffice. Time limitations forbid the extensive consideration of the details of this great field of learning; but it will be quite possible to develop for you, in bold outline, a representation of the system as a whole, that is, to give a sketch of the fundamental laws which rule in the physics of today, of the most important hypotheses employed, and of the great ideas which have recently forced themselves into the subject. I will often gladly endeavor to go into details, but not in the sense of a thorough treatment of the subject, and only with the object of making the general laws more clear, through appropriate specially chosen examples. I shall select these examples from the most varied branches of physics.

If we wish to obtain a correct understanding of the achievements of theoretical physics, we must guard in equal measure against the mistake of overestimating these achievements, and on the other hand, against the corresponding mistake of underestimating them. That the second mistake is actually often made, is shown by the circumstance that quite recently voices have been loudly raised maintaining the bankruptcy and, débâcle of the whole of natural science. But I think such assertions may easily be refuted by reference to the simple fact that with each decade the number and the significance of the means increase, whereby mankind learns directly through the aid of theoretical physics to make nature useful for its own purposes. The technology of today would be impossible without the aid of theoretical physics. The development of the whole of electro-technics from galvanoplasty to wireless telegraphy is a striking proof of this, not to mention aerial navigation. On the other hand, the mistake of overestimating the achievements of theoretical physics appears to me to be much more dangerous, and this danger is particularly threatened by those who have penetrated comparatively little into the heart of the subject. They maintain that some time, through a proper improvement of our science, it will be possible, not only to represent completely through physical formulae the inner constitution of the atoms, but also the laws of mental life. I think that there is nothing in the world entitling us to the one or the other of these expectations. On the other hand, I believe that there is much which directly opposes them. Let us endeavor then to follow the middle course and not to deviate appreciably toward the one side or the other.

When we seek for a solid immovable foundation which is able to carry the whole structure of theoretical physics, we meet with the questions: What lies at the bottom of physics? What is the material with which it operates? Fortunately, there is a complete answer to this question. The material with which theoretical physics operates is measurements, and mathematics is the chief tool with which this material is worked. All physical ideas depend upon measurements, more or less exactly carried out, and each physical definition, each physical law, possesses a more definite significance the nearer it can be brought into accord with the results of measurements. Now measurements are made with the aid of the senses; before all with that of sight, with hearing and with feeling. Thus far, one can say that the origin and the foundation of all physical research are seated in our sense perceptions. Through sense perceptions only do we experience anything of nature; they are the highest court of appeal in questions under dispute. This view is completely confirmed by a glance at the historical development of physical science. Physics grows upon the ground of sensations. The first physical ideas derived were from the individual perceptions of man, and, accordingly, physics was subdivided into: physics of the eye (optics), physics of the ear (acoustics), and physics of heat sensation (theory of heat). It may well be said that so far as there was a domain of sense, so far extended originally the domain of physics. Therefore it appears that in the beginning the division of physics was based upon the peculiarities of man. It possessed, in short, an anthropomorphic character. This appears also, in that physical research, when not occupied with special sense perceptions, is concerned with practical life, and particularly with the practical needs of men. Thus, the art of geodesy led to geometry, the study of machinery to mechanics, and the conclusion lies near that physics in the last analysis had only to do with the sense perceptions and needs of mankind.

In accordance with this view, the sense perceptions are the essential elements of the world; to construct an object as opposed to sense perceptions is more or less an arbitrary matter of will. In fact, when I speak of a tree, I really mean only a complex of sense perceptions: I can see it, I can hear the rustling of its branches, I can smell its fragrance, I experience pain if I knock my head against it, but disregarding all of these sensations, there remains nothing to be made the object of a measurement, wherewith, therefore, natural science can occupy itself. This is certainly true. In accordance with this view, the problem of physics consists only in the relating of sense perceptions, in accordance with experience, to fixed laws; or, as one may express it, in the greatest possible economic accommodation of our ideas to our sensations, an operation which we undertake solely because it is of use to us in the general battle of existence.

All this appears extraordinarily simple and clear and, in accordance with it, the fact may readily be explained that this positivist view is quite widely spread in scientific circles today. It permits, so far as it is limited to the standpoint here depicted (not always done consistently by the exponents of positivism), no hypothesis, no metaphysics; all is clear and plain. I will go still further; this conception never leads to an actual contradiction. I may even say, it can lead to no contradiction. But, ladies and gentlemen, this view has never contributed to any advance in physics. If physics is to advance, in a certain sense its problem must be stated in quite the inverse way, on account of the fact that this conception is inadequate and at bottom possesses only a formal meaning.

The proof of the correctness of this assertion is to be found directly from a consideration of the process of development which theoretical physics has actually undergone, and which one certainly cannot fail to designate as essential. Let us compare the system of physics of today with the earlier and more primitive system which I have depicted above. At the first glance we encounter the most striking difference of all, that in the present system, as well in the division of the various physical domains as in all physical definitions, the historical element plays a much smaller rôle than in the earlier system. While originally, as I have shown above, the fundamental ideas of physics were taken from the specific sense perceptions of man, the latter are today in large measure excluded from physical acoustics, optics, and the theory of heat. The physical definitions of tone, color, and of temperature are today in no wise derived from perception through the corresponding senses; but tone and color are defined through a vibration number or wave length, and the temperature through the volume change of a thermometric substance, or through a temperature scale based on the second law of thermodynamics; but heat sensation is in no wise mentioned in connection with the temperature. With the idea of force it has not been otherwise. Without doubt, the word force originally meant bodily force, corresponding to the circumstance that the oldest tools, the ax, hammer, and mallet, were swung by man's hands, and that the first machines, the lever, roller, and screw, were operated by men or animals. This shows that the idea of force was originally derived from the sense of force, or muscular sense, and was, therefore, a specific sense perception. Consequently, I regard it today as quite essential in a lecture on mechanics to refer, at any rate in the introduction, to the original meaning of the force idea. But in the modern exact definition of force the specific notion of sense perception is eliminated, as in the case of color sense, and we may say, quite in general, that in modern theoretical physics the specific sense perceptions play a much smaller rôle in all physical definitions than formerly. In fact, the crowding into the background of the specific sense elements goes so far that the branches of physics which were originally completely and uniquely characterized by an arrangement in accordance with definite sense perceptions have fallen apart, in consequence of the loosening of the bonds between different and widely separated branches, on account of the general advance towards simplification and coordination. The best example of this is furnished by the theory of heat. Earlier, heat formed a separate and unified domain of physics, characterized through the perceptions of heat sensation. Today one finds in well nigh all physics textbooks dealing with heat a whole domain, that of radiant heat, separated and treated under optics. The significance of heat perception no longer suffices to bring together the heterogeneous parts.

In short, we may say that the characteristic feature of the entire previous development of theoretical physics is a definite elimination from all physical ideas of the anthropomorphic elements, particularly those of specific sense perceptions. On the other hand, as we have seen above, if one reflects that the perceptions form the point of departure in all physical research, and that it is impossible to contemplate their absolute exclusion, because we cannot close the source of all our knowledge, then this conscious departure from the original conceptions must always appear astonishing or even paradoxical. There is scarcely a fact in the history of physics which today stands out so clearly as this. Now, what are the great advantages to be gained through such a real obliteration of personality? What is the result for the sake of whose achievement are sacrificed the directness and succinctness such as only the special sense perceptions vouchsafe to physical ideas?

The result is nothing more than the attainment of unity and compactness in our system of theoretical physics, and, in fact, the unity of the system, not only in relation to all of its details, but also in relation to physicists of all places, all times, all peoples, all cultures. Certainly, the system of theoretical physics should be adequate, not only for the inhabitants of this earth, but also for the inhabitants of other heavenly bodies. Whether the inhabitants of Mars, in case such actually exist, have eyes and ears like our own, we do not know,—it is quite improbable; but that they, in so far as they possess the necessary intelligence, recognize the law of gravitation and the principle of energy, most physicists would hold as self evident: and anyone to whom this is not evident had better not appeal to the physicists, for it will always remain for him an unsolvable riddle that the same physics is made in the United States as in Germany.

To sum up, we may say that the characteristic feature of the actual development of the system of theoretical physics is an ever extending emancipation from the anthropomorphic elements, which has for its object the most complete separation possible of the system of physics and the individual personality of the physicist. One may call this the objectiveness of the system of physics. In order to exclude the possibility of any misunderstanding, I wish to emphasize particularly that we have here to do, not with an absolute separation of physics from the physicist—for a physics without the physicist is unthinkable,—but with the elimination of the individuality of the particular physicist and therefore with the production of a common system of physics for all physicists.

Now, how does this principle agree with the positivist conceptions mentioned above? Separation of the system of physics from the individual personality of the physicist? Opposed to this principle, in accordance with those conceptions, each particular physicist must have his special system of physics, in case that complete elimination of all metaphysical elements is effected; for physics occupies itself only with the facts discovered through perceptions, and only the individual perceptions are directly involved. That other living beings have sensations is, strictly speaking, but a very probable, though arbitrary, conclusion from analogy. The system of physics is therefore primarily an individual matter and, if two physicists accept the same system, it is a very happy circumstance in connection with their personal relationship, but it is not essentially necessary. One can regard this view-point however he will; in physics it is certainly quite fruitless, and this is all that I care to maintain here. Certainly, I might add, each great physical idea means a further advance toward the emancipation from anthropomorphic ideas. This was true in the passage from the Ptolemaic to the Copernican cosmical system, just as it is true at the present time for the apparently impending passage from the so-called classical mechanics of mass points to the general dynamics originating in the principle of relativity. In accordance with this, man and the earth upon which he dwells are removed from the centre of the world. It may be predicted that in this century the idea of time will be divested of the absolute character with which men have been accustomed to endow it (cf. the final lecture). Certainly, the sacrifices demanded by every such revolution in the intuitive point of view are enormous; consequently, the resistance against such a change is very great. But the development of science is not to be permanently halted thereby; on the contrary, its strongest impetus is experienced through precisely those forces which attain success in the struggle against the old points of view, and to this extent such a struggle is constantly necessary and useful.

Now, how far have we advanced today toward the unification of our system of physics? The numerous independent domains of the earlier physics now appear reduced to two; mechanics and electrodynamics, or, as one may say: the physics of material bodies and the physics of the ether. The former comprehends acoustics, phenomena in material bodies, and chemical phenomena; the latter, magnetism, optics, and radiant heat. But is this division a fundamental one? Will it prove final? This is a question of great consequence for the future development of physics. For myself, I believe it must be answered in the negative, and upon the following grounds: mechanics and electrodynamics cannot be permanently sharply differentiated from each other. Does the process of light emission, for example, belong to mechanics or to electrodynamics? To which domain shall be assigned the laws of motion of electrons? At first glance, one may perhaps say: to electrodynamics, since with the electrons ponderable matter does not play any rôle. But let one direct his attention to the motion of free electrons in metals. There he will find, in the study of the classical researches of H. A. Lorentz, for example, that the laws obeyed by the electrons belong rather to the kinetic theory of gases than to electrodynamics. In general, it appears to me that the original differences between processes in the ether and processes in material bodies are to be considered as disappearing. Electrodynamics and mechanics are not so remarkably far apart, as is considered to be the case by many people, who already speak of a conflict between the mechanical and the electrodynamic views of the world. Mechanics requires for its foundation essentially nothing more than the ideas of space, of time, and of that which is moving, whether one considers this as a substance or a state. The same ideas are also involved in electrodynamics. A sufficiently generalized conception of mechanics can therefore also well include electrodynamics, and, in fact, there are many indications pointing toward the ultimate amalgamation of these two subjects, the domains of which already overlap in some measure.

If, therefore, the gulf between ether and matter be once bridged, what is the point of view which in the last analysis will best serve in the subdivision of the system of physics? The answer to this question will characterize the whole nature of the further development of our science. It is, therefore, the most important among all those which I propose to treat today. But for the purposes of a closer investigation it is necessary that we go somewhat more deeply into the peculiarities of physical principles.

We shall best begin at that point from which the first step was made toward the actual realization of the unified system of physics previously postulated by the philosophers only; at the principle of conservation of energy. For the idea of energy is the only one besides those of space and time which is common to all the various domains of physics. In accordance with what I have stated above, it will be apparent and quite self evident to you that the principle of energy, before its general formularization by Mayer, Joule, and Helmholz, also bore an anthropomorphic character. The roots of this principle lay already in the recognition of the fact that no one is able to obtain useful work from nothing; and this recognition had originated essentially in the experiences which were gathered in attempts at the solution of a technical problem: the discovery of perpetual motion. To this extent, perpetual motion has come to have for physics a far reaching significance, similar to that of alchemy for the chemist, although it was not the positive, but rather the negative results of these experiments, through which science was advanced. Today we speak of the principle of energy quite without reference to the technical viewpoint or to that of man. We say that the total amount of energy of an isolated system of bodies is a quantity whose amount can be neither increased nor diminished through any kind of process within the system, and we no longer consider the accuracy with which this law holds as dependent upon the refinement of the methods, which we at present possess, of testing experimentally the question of the realization of perpetual motion. In this, strictly speaking, unprovable generalization, impressed upon us with elemental force, lies the emancipation from the anthropomorphic elements mentioned above.

While the principle of energy stands before us as a complete independent structure, freed from and independent of the accidents appertaining to its historical development, this is by no means true in equal measure in the case of that principle which R. Clausius introduced into physics; namely, the second law of thermodynamics. This law plays a very peculiar rôle in the development of physical science, to the extent that one is not able to assert today that for it a generally recognized, and therefore objective formularization, has been found. In our present consideration it is therefore a matter of particular interest to examine more closely its significance.

In contrast to the first law of thermodynamics, or the energy principle, the second law may be characterized as follows. While the first law permits in all processes of nature neither the creation nor destruction of energy, but permits of transformations only, the second law goes still further into the limitation of the possible processes of nature, in that it permits, not all kinds of transformations, but only certain types, subject to certain conditions. The second law occupies itself, therefore, with the question of the kind and, in particular, with the direction of any natural process.

At this point a mistake has frequently been made, which has hindered in a very pronounced manner the advance of science up to the present day. In the endeavor to give to the second law of thermodynamics the most general character possible, it has been proclaimed by followers of W. Ostwald as the second law of energetics, and the attempt made so to formulate it that it shall determine quite generally the direction of every process occurring in nature. Some weeks ago I read in a public academic address of an esteemed colleague the statement that the import of the second law consists in this, that a stone falls downwards, that water flows not up hill, but down, that electricity flows from a higher to a lower potential, and so on. This is a mistake which at present is altogether too prevalent not to warrant mention here.

The truth is, these statements are false. A stone can just as well rise in the air as fall downwards; water can likewise flow upwards, as, for example, in a spring; electricity can flow very well from a lower to a higher potential, as in the case of oscillating discharge of a condenser. The statements are obviously quite correct, if one applies them to a stone originally at rest, to water at rest, to electricity at rest; but then they follow immediately from the energy principle, and one does not need to add a special second law. For, in accordance with the energy principle, the kinetic energy of the stone or of the water can only originate at the cost of gravitational energy, i. e., the center of mass must descend. If, therefore, motion is to take place at all, it is necessary that the gravitational energy shall decrease. That is, the center of mass must descend. In like manner, an electric current between two condenser plates can originate only at the cost of electrical energy already present; the electricity must therefore pass to a lower potential. If, however, motion and current be already present, then one is not able to say, a priori, anything in regard to the direction of the change; it can take place just as well in one direction as the other. Therefore, there is no new insight into nature to be obtained from this point of view.

Upon an equally inadequate basis rests another conception of the second law, which I shall now mention. In considering the circumstance that mechanical work may very easily be transformed into heat, as by friction, while on the other hand heat can only with difficulty be transformed into work, the attempt has been made so to characterize the second law, that in nature the transformation of work into heat can take place completely, while that of heat into work, on the other hand, only incompletely and in such manner that every time a quantity of heat is transformed into work another corresponding quantity of energy must necessarily undergo at the same time a compensating transformation, as, e. g., the passage of heat from a higher to a lower temperature. This assertion is in certain special cases correct, but does not strike in general at the true import of the matter, as I shall show by a simple example.

One of the most important laws of thermodynamics is, that the total energy of an ideal gas depends only upon its temperature, and not upon its volume. If an ideal gas be allowed to expand while doing work, and if the cooling of the gas be prevented through the simultaneous addition of heat from a heat reservoir at higher temperature, the gas remains unchanged in temperature and energy content, and one may say that the heat furnished by the heat reservoir is completely transformed into work without exchange of energy. Not the least objection can be urged against this assertion. The law of incomplete transformation of heat into work is retained only through the adoption of a different point of view, but which has nothing to do with the status of the physical facts and only modifies the way of looking at the matter, and therefore can neither be supported nor contradicted through facts; namely, through the introduction ad hoc of new particular kinds of energy, in that one divides the energy of the gas into numerous parts which individually can depend upon the volume. But it is a priori evident that one can never derive from so artificial a definition a new physical law, and it is with such that we have to do when we pass from the first law, the principle of conservation of energy, to the second law.

I desire now to introduce such a new physical law: "It is not possible to construct a periodically functioning motor which in principle does not involve more than the raising of a load and the cooling of a heat reservoir." It is to be understood, that in one cycle of the motor quite arbitrary complicated processes may take place, but that after the completion of one cycle there shall remain no other changes in the surroundings than that the heat reservoir is cooled and that the load is raised a corresponding distance, which may be calculated from the first law. Such a motor could of course be used at the same time as a refrigerating machine also, without any further expenditure of energy and materials. Such a motor would moreover be the most efficient in the world, since it would involve no cost to run it; for the earth, the atmosphere, or the ocean could be utilized as the heat reservoir. We shall call this, in accordance with the proposal of W. Ostwald, perpetual motion of the second kind. Whether in nature such a motion is actually possible cannot be inferred from the energy principle, and may only be determined by special experiments.

Just as the impossibility of perpetual motion of the first kind leads to the principle of the conservation of energy, the quite independent principle of the impossibility of perpetual motion of the second kind leads to the second law of thermodynamics, and, if we assume this impossibility as proven experimentally, the general law follows immediately: *there are processes in nature which in no possible way can be made completely reversible.* For consider, e. g., a frictional process through which mechanical work is transformed into heat with the aid of suitable apparatus, if it were actually possible to make in some way such complicated apparatus completely reversible, so that everywhere in nature exactly the same conditions be reestablished as existed at the beginning of the frictional process, then the apparatus considered would be nothing more than the motor described above, furnishing a perpetual motion of the second kind. This appears evident immediately, if one clearly perceives what the apparatus would accomplish: transformation of heat into work without any further outstanding change.

We call such a process, which in no wise can be made completely reversible, an irreversible process, and all other processes reversible processes; and thus we strike the kernel of the second law of thermodynamics when we say that irreversible processes occur in nature. In accordance with this, the changes in nature have a unidirectional tendency. With each irreversible process the world takes a step forward, the traces of which under no circumstances can be completely obliterated. Besides friction, examples of irreversible processes are: heat conduction, diffusion, conduction of electricity in conductors of finite resistance, emission of light and heat radiation, disintegration of the atom in radioactive substances, and so on. On the other hand, examples of reversible processes are: motion of the planets, free fall in empty space, the undamped motion of a pendulum, the frictionless flow of liquids, the propagation of light and sound waves without absorption and refraction, undamped electrical vibrations, and so on. For all these processes are already periodic or may be made completely reversible through suitable contrivances, so that there remains no outstanding change in nature; for example, the free fall of a body whereby the acquired velocity is utilized to raise the body again to its original height; a light or sound wave which is allowed in a suitable manner to be totally reflected from a perfect mirror.

What now are the general properties and criteria of irreversible processes, and what is the general quantitative measure of irreversibility? This question has been examined and answered in the most widely different ways, and it is evident here again how difficult it is to reach a correct formularization of a problem. Just as originally we came upon the trail of the energy principle through the technical problem of perpetual motion, so again a technical problem, namely, that of the steam engine, led to the differentiation between reversible and irreversible processes. Long ago Sadi Carnot recognized, although he utilized an incorrect conception of the nature of heat, that irreversible processes are less economical than reversible, or that in an irreversible process a certain opportunity to derive mechanical work from heat is lost. What then could have been simpler than the thought of making, quite in general, the measure of the irreversibility of a process the quantity of mechanical work which is unavoidably lost in the process. For a reversible process then, the unavoidably lost work is naturally to be set equal to zero. This view, in accordance with which the import of the second law consists in a dissipation of useful energy, has in fact, in certain special cases, e. g., in isothermal processes, proved itself useful. It has persisted, therefore, in certain of its aspects up to the present day; but for the general case, however, it has shown itself as fruitless and, in fact, misleading. The reason for this lies in the fact that the question concerning the lost work in a given irreversible process is by no means to be answered in a determinate manner, so long as nothing further is specified with regard to the source of energy from which the work considered shall be obtained.

An example will make this clear. Heat conduction is an irreversible process, or as Clausius expresses it: Heat cannot without compensation pass from a colder to a warmer body. What now is the work which in accordance with definition is lost when the quantity of heat Q passes through direct conduction from a warmer body at the temperature T_1 to a colder body at the temperature T_2? In order to answer this question, we make use of the heat transfer involved in carrying out a reversible Carnot cyclical process between the two bodies employed as heat reservoirs. In this process a certain amount of work would be obtained, and it is just the amount sought, since it is that which would be lost in the direct passage by conduction; but this has no definite value so long as we do not know whence the work originates, whether, e. g., in the warmer body or in the colder body, or from somewhere else. Let one reflect that the heat given up by the warmer body in the reversible process is certainly not equal to the heat absorbed by the colder body, because a certain amount of heat is transformed into work, and that we can identify, with exactly the same right, the quantity of heat Q transferred by the direct process of conduction with that which in the cyclical process is given up by the warmer body, or with that absorbed by the colder body. As one does the former or the latter, he accordingly obtains for the quantity of lost work in the process of conduction:

$$Q \cdot \frac{T_1 - T_2}{T_1} \quad \text{or} \quad Q \cdot \frac{T_1 - T_2}{T_2}.$$

We see, therefore, that the proposed method of expressing mathematically the irreversibility of a process does not in general effect its object, and at the same time we recognize the peculiar reason which prevents its doing so. The statement of the question is too anthropomorphic. It is primarily too much concerned with the needs of mankind, in that it refers directly to the acquirement of useful work. If one require from nature a determinate answer, he must take a more general point of view, more disinterested, less economic. We shall now seek to do this.

Let us consider any typical process occurring in nature. This will carry all bodies concerned in it from a determinate initial state, which I designate as state A, into a determinate final state B. The process is either reversible or irreversible. A third possibility is excluded. But whether it is reversible or irreversible depends solely upon the nature of the two states A and B, and not at all upon the way in which the process has been carried out; for we are only concerned with the answer to the question as to whether or not, when the state B is once reached, a complete return to A in any conceivable manner may be accomplished. If now, the complete return from B to A is not possible, and the process therefore irreversible, it is obvious that the state B may be distinguished in nature through a certain property from state A. Several years ago I ventured to express this as follows: that nature possesses a greater "preference" for state B than for state A. In accordance with this mode of expression, all those processes of nature are impossible for whose final state nature possesses a smaller preference than for the original state. Reversible processes constitute a limiting case; for such, nature possesses an equal preference for the initial and for the final state, and the passage between them takes place as well in one direction as the other.

We have now to seek a physical quantity whose magnitude shall serve as a general measure of the preference of nature for a given state. This quantity must be one which is directly determined by the state of the system considered, without reference to the previous history of the system, as is the case with the energy, with the volume, and with other properties of the system. It should possess the peculiarity of increasing in all irreversible processes and of remaining unchanged in all reversible processes, and the amount of change which it experiences in a process would furnish a general measure for the irreversibility of the process.

R. Clausius actually found this quantity and called it "entropy." Every system of bodies possesses in each of its states a definite entropy, and this entropy expresses the preference of nature for the state in question. It can, in all the processes which take place within the system, only increase and never decrease. If it be desired to consider a process in which external actions upon the system are present, it is necessary to consider those bodies in which these actions originate as constituting part of the system; then the law as stated in the above form is valid. In accordance with it, the entropy of a system of bodies is simply equal to the sum of the entropies of the individual bodies, and the entropy of a single body is, in accordance with Clausius, found by the aid of a certain reversible process. Conduction of heat to a body increases its entropy, and, in fact, by an amount equal to the ratio of the quantity of heat given the body to its temperature. Simple compression, on the other hand, does not change the entropy.

Returning to the example mentioned above, in which the

quantity of heat Q is conducted from a warmer body at the temperature T_1 to a colder body at the temperature T_2, in accordance with what precedes, the entropy of the warmer body decreases in this process, while, on the other hand, that of the colder increases, and the sum of both changes, that is, the change of the total entropy of both bodies, is:

$$-\frac{Q}{T_1}+\frac{Q}{T_2}>0.$$

This positive quantity furnishes, in a manner free from all arbitrary assumptions, the measure of the irreversibility of the process of heat conduction. Such examples may be cited indefinitely. Every chemical process furnishes an increase of entropy.

We shall here consider only the most general case treated by Clausius: an arbitrary reversible or irreversible cyclical process, carried out with any physico-chemical arrangement, utilizing an arbitrary number of heat reservoirs. Since the arrangement at the conclusion of the cyclical process is the same as that at the beginning, the final state of the process is to be distinguished from the initial state solely through the different heat content of the heat reservoirs, and in that a certain amount of mechanical work has been furnished or consumed. Let Q be the heat given up in the course of the process by a heat reservoir at the temperature T, and let A be the total work yielded (consisting, e. g., in the raising of weights); then, in accordance with the first law of thermodynamics:

$$\Sigma Q = A.$$

In accordance with the second law, the sum of the changes in entropy of all the heat reservoirs is positive, or zero. It follows, therefore, since the entropy of a reservoir is decreased by the amount Q/T through the loss of heat Q that:

$$\Sigma\frac{Q}{T}\leqq 0.$$

This is the well-known inequality of Clausius.

In an isothermal cyclical process, T is the same for all reservoirs. Therefore:

$$\Sigma Q \leqq 0, \quad \text{hence: } A \leqq 0.$$

That is: in an isothermal cyclical process, heat is produced and work is consumed. In the limiting case, a reversible isothermal cyclical process, the sign of equality holds, and therefore the work consumed is zero, and also the heat produced. This law plays a leading rôle in the application of thermodynamics to physical chemistry.

The second law of thermodynamics including all of its consequences has thus led to the principle of increase of entropy. You will now readily understand, having regard to the questions mentioned above, why I express it as my opinion that in the theoretical physics of the future the first and most important differentiation of all physical processes will be into reversible and irreversible processes.

In fact, all reversible processes, whether they take place in material bodies, in the ether, or in both together, show a much greater similarity among themselves than to any irreversible process. In the differential equations of reversible processes the time differential enters only as an even power, corresponding to the circumstance that the sign of time can be reversed. This holds equally well for vibrations of the pendulum, electrical vibrations, acoustic and optical waves, and for motions of mass points or of electrons, if we only exclude every kind of damping. But to such processes also belong those infinitely slow processes of thermodynamics which consist of states of equilibrium in which the time in general plays no rôle, or, as one may also say, occurs with the zero power, which is to be reckoned as an even power. As Helmholtz has pointed out, all these reversible processes have the common property that they may be completely represented by the principle of least action, which gives a definite answer to all questions concerning any such measurable process, and, to this extent, theory of reversible processes may be regarded as completely established. Reversible processes have, however, the disadvantage that singly and collectively they are only ideal: in actual nature there is no such thing as a reversible process. Every natural process involves in greater or less degree friction or conduction of heat. But in the domain of irreversible processes the principle of least action is no longer sufficient; for the principle of increase of entropy brings into the system of physics a wholly new element, foreign to the action principle, and which demands special mathematical treatment. The unidirectional course of a process in the attainment of a fixed final state is related to it.

I hope the foregoi ıg considerations have sufficed to make clear to you that the distinction between reversible and irreversible processes is much broader than that between mechanical and electrical processes and that, therefore, this difference, with better right than any other, may be taken advantage of in classifying all physical processes, and that it may eventually play in the theoretical physics of the future the principal rôle.

However, the classification mentioned is in need of quite an essential improvement, for it cannot be denied that in the form set forth, the system of physics is still suffering from a strong dose of anthropomorphism. In the definition of irreversibility, as well as in that of entropy, reference is made to the possibility of carrying out in nature certain changes, and this means, fundamentally, nothing more than that the division of physical processes is made dependent upon the manipulative skill of man in the art of experimentation, which certainly does not always remain at a fixed stage, but is continually being more and more perfected. If, therefore, the distinction between reversible and irreversible processes is actually to have a lasting significance for all times, it must be essentially broadened and made independent of any reference to the capacities of mankind. How this may happen, I desire to state one week from tomorrow. The lecture of tomorrow will be devoted to the problem of bringing before you some of the most important of the great number of practical consequences following from the entropy principle.

SECOND LECTURE.

Thermodynamic States of Equilibrium in Dilute Solutions.

In the lecture of yesterday I sought to make clear the fact that the essential, and therefore the final division of all processes occurring in nature, is into reversible and irreversible processes, and the characteristic difference between these two kinds of processes, as I have further separated them, is that in irreversible processes the entropy increases, while in all reversible processes it remains constant. Today I am constrained to speak of some of the consequences of this law which will illustrate its rich fruitfulness. They have to do with the question of the laws of thermodynamic equilibrium. Since in nature the entropy can only increase, it follows that the state of a physical configuration which is completely isolated, and in which the entropy of the system possesses an absolute maximum, is necessarily a state of stable equilibrium, since for it no further change is possible. How deeply this law underlies all physical and chemical relations has been shown by no one better and more completely than by John Willard Gibbs, whose name, not only in America, but in the whole world will be counted among those of the most famous theoretical physicists of all times; to whom, to my sorrow, it is no longer possible for me to tender personally my respects. It would be gratuitous for me, here in the land of his activity, to expatiate fully on the progress of his ideas, but you will perhaps permit me to speak in the lecture of today of some of the important applications in which thermodynamic research, based on Gibbs works, can be advanced beyond his results.

These applications refer to the theory of dilute solutions, and we shall occupy ourselves today with these, while I show you by a definite example what fruitfulness is inherent in thermodynamic theory. I shall first characterize the problem quite generally. It has to do with the state of equilibrium of a material system of any number of arbitrary constituents in an arbitrary number of phases, at a given temperature T and given pressure p. If the system is completely isolated, and therefore guarded against all external thermal and mechanical actions, then in any ensuing change the entropy of the system will increase:

$$dS > 0.$$

But if, as we assume, the system stands in such relation to its surroundings that in any change which the system undergoes the temperature T and the pressure p are maintained constant, as, for instance, through its introduction into a calorimeter of great heat capacity and through loading with a piston of fixed weight, the inequality would suffer a change thereby. We must then take account of the fact that the surrounding bodies also, e. g., the calorimetric liquid, will be involved in the change. If we denote the entropy of the surrounding bodies by S_0, then the following more general equation holds:

$$dS + dS_0 > 0.$$

In this equation

$$dS_0 = -\frac{Q}{T},$$

if Q denote the heat which is given up in the change by the surroundings to the system. On the other hand, if U denote the energy, V the volume of the system, then, in accordance with the first law of thermodynamics,

$$Q = dU + pdV.$$

Consequently, through substitution:

$$dS - \frac{dU + pdV}{T} > 0$$

or, since p and T are constant:

$$d\left(S - \frac{U + pV}{T}\right) > 0.$$

If, therefore, we put:

$$S - \frac{U + pV}{T} = \Phi, \tag{1}$$

then

$$d\Phi > 0,$$

and we have the general law, that in every isothermal-isobaric (T = const., p = const.) change of state of a physical system the quantity Φ increases. The absolutely stable state of equilibrium of the system is therefore characterized through the maximum of Φ:

$$\delta\Phi = 0. \tag{2}$$

If the system consist of numerous phases, then, because Φ, in accordance with (1), is linear and homogeneous in S, U and V, the quantity Φ referring to the whole system is the sum of the quantities Φ referring to the individual phases. If the expression for Φ is known as a function of the independent variables for each phase of the system, then, from equation (2), all questions concerning the conditions of stable equilibrium may be answered. Now, within limits, this is the case for dilute solutions. By "solution" in thermodynamics is meant each homogeneous phase, in whatever state of aggregation, which is composed of a series of different molecular complexes, each of which is represented by a definite molecular number. If the molecular number of a given complex is great with reference to all the remaining complexes, then the solution is called dilute, and the molecular complex in question is called the solvent; the remaining complexes are called the dissolved substances.

Let us now consider a dilute solution whose state is determined by the pressure p, the temperature T, and the molecular numbers $n_0, n_1, n_2, n_3, \cdots$, wherein the subscript zero refers to the solvent. Then the numbers $n_1, n_2, n_3, \cdots$ are all small with respect to n_0, and on this account the volume V and the energy U are linear functions of the molecular numbers:

$$V = n_0 v_0 + n_1 v_1 + n_2 v_2 + \cdots,$$

$$U = n_0 u_0 + n_1 u_1 + n_2 u_2 + \cdots,$$

wherein the v's and u's depend upon p and T only.

From the general equation of entropy:

$$dS = \frac{dU + pdV}{T},$$

in which the differentials depend only upon changes in p and T, and not in the molecular numbers, there results therefore:

$$dS = n_0 \frac{du_0 + pdv_0}{T} + n_1 \frac{du_1 + pdv_1}{T} + \cdots,$$

and from this it follows that the expressions multiplied by $n_0, n_1 \cdots$, dependent upon p and T only, are complete differentials. We may therefore write:

$$\frac{du_0 + pdv_0}{T} = ds_0, \quad \frac{du_1 + pdv_1}{T} = ds_1, \quad \cdots \tag{3}$$

and by integration obtain:

$$S = n_0 s_0 + n_1 s_1 + n_2 s_2 + \cdots + C.$$

The constant C of integration does not depend upon p and T, but may depend upon the molecular numbers $n_0, n_1, n_2, \cdots$. In order to express this dependence generally, it suffices to know it for a special case, for fixed values of p and T. Now every solution passes, through appropriate increase of temperature and decrease of pressure, into the state of a mixture of ideal gases, and for this case the entropy is fully known, the integration constant being, in accordance with Gibbs:

$$C = -R(n_0 \log c_0 + n_1 \log c_1 + \cdots),$$

wherein R denotes the absolute gas constant and $c_0, c_1, c_2, \cdots$

denote the "molecular concentrations":

$$c_0 = \frac{n_0}{n_0 + n_1 + n_2 + \cdots}, \quad c_1 = \frac{n_1}{n_0 + n_1 + n_2 + \cdots}, \quad \cdots.$$

Consequently, quite in general, the entropy of a dilute solution is:

$$S = n_0(s_0 - R \log c_0) + n_1(s_1 - R \log c_1) + \cdots,$$

and, finally, from this it follows by substitution in equation (1) that:

$$\Phi = n_0(\varphi_0 - R \log c_0) + n_1(\varphi_1 - R \log c_1) + \cdots, \qquad (4)$$

if we put for brevity:

$$\varphi_0 = s_0 - \frac{u_0 + pv_0}{T}, \quad \varphi_1 = s_1 - \frac{u_1 + pv_1}{T}, \quad \cdots \qquad (5)$$

all of which quantities depend only upon p and T.

With the aid of the expression obtained for Φ we are enabled through equation (2) to answer the question with regard to thermodynamic equilibrium. We shall first find the general law of equilibrium and then apply it to a series of particularly interesting special cases.

Every material system consisting of an arbitrary number of homogeneous phases may be represented symbolically in the following way:

$$n_0 m_0,\ n_1 m_1,\ \cdots \mid n_0' m_0',\ n_1' m_1',\ \cdots \mid n_0'' m_0'',\ n_1'' m_1'',\ \cdots \mid \cdots.$$

Here the molecular numbers are denoted by n, the molecular weights by m, and the individual phases are separated from one another by vertical lines. We shall now suppose that each phase represents a dilute solution. This will be the case when each phase contains only a single molecular complex and therefore represents an absolutely pure substance; for then the concentrations of all the dissolved substances will be zero.

If now an isobaric-isothermal change in the system of such kind is possible that the molecular numbers

$$n_0,\ n_1,\ n_2,\ \cdots, \quad n_0',\ n_1',\ n_2',\ \cdots, \quad n_0'',\ n_1'',\ n_2'',\ \cdots$$

change simultaneously by the amounts

$$\delta n_0, \delta n_1, \delta n_2, \cdots, \quad \delta n_0', \delta n_1', \delta n_2', \cdots, \quad \delta n_0'', \delta n_1'', \delta n_2'', \cdots$$

then, in accordance with equation (2), equilibrium obtains with respect to the occurrence of this change if, when T and p are held constant, the function

$$\Phi + \Phi' + \Phi'' + \cdots$$

is a maximum, or, in accordance with equation (4):

$$\Sigma(\varphi_0 - R \log c_0)\delta n_0 + (\varphi_1 - R \log c_1)\delta n_1 + \cdots = 0$$

(the summation Σ being extended over all phases of the system). Since we are only concerned in this equation with the ratios of the δn's, we put

$$\delta n_0 : \delta n_1 : \cdots : \delta n_0' : \delta n_1' . \cdots : \delta n_0'' : \delta n_1'' : \cdots$$
$$= \nu_0 : \nu_1 : \cdots : \nu_0' : \nu_1' : \cdots : \nu_0'' : \nu_1'' : \cdots,$$

wherein we are to understand by the simultaneously changing ν's, in the variation considered, simple integer positive or negative numbers, according as the molecular complex under consideration is formed or disappears in the change. Then the condition for equilibrium is:

$$\Sigma\nu_0 \log c_1 + \nu_1 \log c_1 + \cdots = \frac{1}{R}\Sigma\nu_0\varphi_0 + \nu_1\varphi_1 + \cdots = \log K. \quad (6)$$

K and the quantities $\varphi_0, \varphi_1, \varphi_2, \cdots$ depend only upon p and T, and this dependence is to be found from the equations:

$$\frac{\partial \log K}{\partial p} = \frac{1}{R}\Sigma\nu_0 \frac{\partial \varphi_0}{\partial p} + \nu_1 \frac{\partial \varphi_1}{\partial p} + \cdots,$$

$$\frac{\partial \log K}{\partial T} = \frac{1}{R}\Sigma\nu_0 \frac{\partial \varphi_0}{\partial T} + \nu_1 \frac{\partial \varphi_1}{\partial T} + \cdots.$$

Now, in accordance with (5), for any infinitely small change of p and T:

$$d\varphi_0 = ds_0 - \frac{du_0 + pdv_0 + v_0dp}{T} + \frac{u_0 + pv_0}{T^2} \cdot dT,$$

and consequently, from (3):

$$d\varphi_0 = \frac{u_0 + pv_0}{T^2} dT - \frac{v_0 dp}{T},$$

and hence:

$$\frac{\partial \varphi_0}{\partial p} = -\frac{v_0}{T}, \quad \frac{\partial \varphi_0}{\partial T} = \frac{u_0 + pv_0}{T^2}.$$

Similar equations hold for the other φ's, and therefore we get:

$$\frac{\partial \log K}{\partial p} = -\frac{1}{RT} \Sigma \nu_0 v_0 + \nu_1 v_1 + \cdots,$$

$$\frac{\partial \log K}{\partial T} = -\frac{1}{RT^2} \Sigma \nu_0 u_0 + \nu_2 u_2 + \cdots + p(\nu_0 v_0 + \nu_1 v_1 + \cdots)$$

or, more briefly:

$$\frac{\partial \log K}{\partial p} = -\frac{1}{RT} \cdot \Delta V, \quad \frac{\partial \log K}{\partial T} = \frac{\Delta Q}{RT^2}, \tag{7}$$

if ΔV denote the change in the total volume of the system and ΔQ the heat which is communicated to it from outside, during the isobaric isothermal change considered. We shall now investigate the import of these relations in a series of important applications.

I. *Electrolytic Dissociation of Water.*

The system consists of a single phase:

$$n_0 H_2O, \ n_1 \overset{+}{H}, \ n_2 \overset{-}{HO}.$$

The transformation under consideration

$$\nu_0 : \nu_1 : \nu_2 = \delta n_0 : \delta n_1 : \delta n_2$$

consists in the dissociation of a molecule H_2O into a molecule $\overset{+}{H}$ and a molecule $\overset{-}{HO}$, therefore:

$$\nu_0 = -1, \quad \nu_1 = 1, \quad \nu_2 = 1.$$

Hence, in accordance with (6), for equilibrium:

$$-\log c_0 + \log c_1 + \log c_2 = \log K,$$

or, since $c_1 = c_2$ and $c_0 = 1$, approximately:

$$2 \log c_1 = \log K.$$

The dependence of the concentration c_1 upon the temperature now follows from (7):

$$2 \frac{\partial \log c_1}{\partial T} = \frac{\Delta Q}{RT^2}.$$

ΔQ, the quantity of heat which it is necessary to supply for the dissociation of a molecule of H_2O into the ions $\overset{+}{H}$ and $H\bar{O}$, is, in accordance with Arrhenius, equal to the heat of ionization in the neutralization of a strong univalent base and acid in a dilute aqueous solution, and, therefore, in accordance with the recent measurements of Wörmann,[1]

$$\Delta Q = 27{,}857 - 48.5T \text{ gr. cal.}$$

Using the number 1.985 for the ratio of the absolute gas constant R to the mechanical equivalent of heat, it follows that:

$$\frac{\partial \log c_1}{\partial T} = \frac{1}{2 \cdot 1.985}\left(\frac{27{,}857}{T^2} - \frac{48.5}{T}\right),$$

and by integration:

$$\overset{10}{\log}\, c_1 = -\frac{3047.3}{T} - 12.125 \overset{10}{\log}\, T + \text{const.}$$

This dependence of the degree of dissociation upon the temperature agrees very well with the measurements of the electric conductivity of water at different temperatures by Kohlrausch and Heydweiller, Noyes, and Lundén.

II. *Dissociation of a Dissolved Electrolyte.*

Let the system consists of an aqueous solution of acetic acid:

$$n_0 H_2O, \; n_1 H_4C_2O_2, \; n_2 \overset{+}{H}, \; n_3 H_3\bar{C}_2O_2.$$

The change under consideration consists in the dissociation of a

[1] Ad Heydweiller, Ann. d. Phys., 28, 506, 1909.

molecule $H_4C_2O_2$ into its two ions, therefore

$$\nu_0 = 0, \quad \nu_1 = -1, \quad \nu_2 = 1, \quad \nu_3 = 1.$$

Hence, for the state of equilibrium, in accordance with (6):

$$-\log c_1 + \log c_2 + \log c_3 = \log K,$$

or, since $c_2 = c_3$:

$$\frac{c_2^2}{c_1} = K.$$

Now the sum $c_1 + c_2 = c$ is to be regarded as known, since the total number of the undissociated and dissociated acid molecules is independent of the degree of dissociation. Therefore c_1 and c_2 may be calculated from K and c. An experimental test of the equation of equilibrium is possible on account of the connection between the degree of dissociation and electrical conductivity of the solution. In accordance with the electrolytic dissociation theory of Arrhenius, the ratio of the molecular conductivity λ of the solution in any dilution to the molecular conductivity λ_∞ of the solution in infinite dilution is:

$$\frac{\lambda}{\lambda_\infty} = \frac{c_2}{c_1 + c_2} = \frac{c_2}{c},$$

since electric conduction is accounted for by the dissociated molecules only. It follows then, with the aid of the last equation, that:

$$\frac{\lambda^2 c}{\lambda_\infty - \lambda} = K \cdot \lambda_\infty = \text{const.}$$

With unlimited decreasing c, λ increases to λ_∞. This "law of dilution" for binary electrolytes, first enunciated by Ostwald, has been confirmed in numerous cases by experiment, as in the case of acetic acid.

Also, the dependence of the degree of dissociation upon the temperature is indicated here in quite an analogous manner to that in the example considered above, of the dissociation of water.

III. *Vaporization or Solidification of a Pure Liquid.*

In equilibrium the system consists of two phases, one liquid, and one gaseous or solid:

$$n_0 m_0 \mid n_0' m_0'.$$

Each phase contains only a single molecular complex (the solvent), but the molecules in both phases do not need to be the same. Now, if a liquid molecule evaporates or solidifies, then in our notation

$$\nu_0 = -1, \quad \nu_0' = \frac{m_0}{m_0'}, \quad c_0 = 1, \quad c_0' = 1,$$

and consequently the condition for equilibrium, in accordance with (6), is:

$$0 = \log K. \tag{8}$$

Since K depends only upon p and T, this equation therefore expresses a definite relation between p and T: the law of dependence of the pressure of vaporization (or melting pressure) upon the temperature, or vice versa. The import of this law is obtained through the consideration of the dependence of the quantity K upon p and T. If we form the complete differential of the last equation, there results:

$$0 = \frac{\partial \log K}{\partial p} dp + \frac{\partial \log K}{\partial T} dT,$$

or, in accordance with (7):

$$0 = -\frac{\Delta V}{T} dp + \frac{\Delta Q}{T^2} dT.$$

If v_0 and v_0' denote the molecular volumes of the two phases, then:

$$\Delta V = \frac{m_0 v_0'}{m_0'} - v_0,$$

consequently:

$$\Delta Q = T\left(\frac{m_0 v_0'}{m_0'} - v_0\right)\frac{dp}{dT},$$

or, referred to unit mass:

$$\frac{\Delta Q}{m_0} = T\left(\frac{v_0'}{m_0'} - \frac{v_0}{m_0}\right) \cdot \frac{dp}{dT},$$

the well-known formula of Carnot and Clapeyron.

IV. *The Vaporization or Solidification of a Solution of Non-Volatile Substances.*

Most aqueous salt solutions afford examples. The symbol of the system in this case is, since the second phase (gaseous or solid) contains only a single molecular complex:

$$n_0 m_0,\ n_1 m_1,\ n_2 m_2,\ \cdots \mid n_0' m_0'.$$

The change is represented by:

$$\nu_0 = -1,\ \nu_1 = 0,\ \nu_2 = 0,\ \cdots \nu_0' = \frac{m_0}{m_0'},$$

and hence the condition of equilibrium, in accordance with (6), is:

$$-\log c_0 = \log K,$$

or, since to small quantities of higher order:

$$c_0 = \frac{n_0}{n_0 + n_1 + n_2 + \cdots} = 1 - \frac{n_1 + n_2 + \cdots}{n_0},$$

$$\frac{n_1 + n_2 + \cdots}{n_0} = \log K. \tag{9}$$

A comparison with formula (8), found in example III, shows that through the solution of a foreign substance there is involved in the total concentration a small proportionate departure from the law of vaporization or solidification which holds for the pure solvent. One can express this, either by saying: at a fixed pressure p, the boiling point or the freezing point T of the solution is different than that (T_0) for the pure solvent, or: at a fixed pressure T the vapor pressure or solidification pressure p of the solution is different from that (p_0) of the pure solvent. Let us calculate the departure in both cases.

1. If T_0 be the boiling (or freezing temperature) of the pure solvent at the pressure p, then, in accordance with (8):

$$(\log K)_{T=T_0} = 0,$$

and by subtraction of (9) there results:

$$\log K - (\log K)_{T=T_0} = \frac{n_1 + n_2 + \cdots}{n_0}.$$

Now, since T is little different from T_0, we may write in place of this equation, with the aid of (7):

$$\frac{\partial \log K}{\partial T}(T - T_0) = \frac{\Delta Q}{RT_0^2}(T - T_0) = \frac{n_1 + n_2 + \cdots}{n_0},$$

and from this it follows that:

$$T - T_0 = \frac{n_1 + n_2 + \cdots}{n_0} \cdot \frac{RT_0^2}{\Delta Q}. \tag{10}$$

This is the law for the raising of the boiling point or for the lowering of the freezing point, first derived by van't Hoff: in the case of freezing ΔQ (the heat taken from the surroundings during the freezing of a liquid molecule) is negative. Since n_0 and ΔQ occur only as a product, it is not possible to infer anything from this formula with regard to the molecular number of the liquid solvent.

2. If p_0 be the vapor pressure of the pure solvent at the temperature T, then, in accordance with (8):

$$(\log K)_{p=p_0} = 0,$$

and by subtraction of (9) there results:

$$\log K - (\log K)_{p=p_0} = \frac{n_1 + n_2 + \cdots}{n_0}.$$

Now, since p and p_0 are nearly equal, with the aid of (7) we may write:

$$\frac{\partial \log K}{\partial p}(p - p_0) = -\frac{\Delta V}{RT}(p - p_0) = \frac{n_1 + n_2 + \cdots}{n_0},$$

and from this it follows, if ΔV be placed equal to the volume of the gaseous molecule produced in the vaporization of a liquid molecule:

$$\Delta V = \frac{m_0}{m_0'}\frac{RT}{p},$$

$$\frac{p_0 - p}{p} = \frac{m_0'}{m_0}\cdot\frac{n_1 + n_2 + \cdots}{n_0}.$$

This is the law of relative depression of the vapor pressure, first derived by van't Hoff. Since n_0 and m_0 occur only as a product, it is not possible to infer from this formula anything with regard to the molecular weight of the liquid solvent. Frequently the factor m_0'/m_0 is left out in this formula; but this is not allowable when m_0 and m_0' are unequal (as, e. g., in the case of water).

V. *Vaporization of a Solution of Volatile Substances.*

(E. g., a Sufficiently Dilute Solution of Propyl Alcohol in Water.)

The system, consisting of two phases, is represented by the following symbol:

$$n_0m_0,\ n_1m_1,\ n_2m_2,\ \cdots \mid n_0'm_0',\ n_1'm_1',\ n_2'm_2',\ \cdots,$$

wherein, as above, the figure 0 refers to the solvent and the figures 1, 2, 3 $\cdots$ refer to the various molecular complexes of the dissolved substances. By the addition of primes in the case of the molecular weights (m_0', m_1', m_2' $\cdots$) the possibility is left open that the various molecular complexes in the vapor may possess a different molecular weight than in the liquid.

Since the system here considered may experience various sorts of changes, there are also various conditions of equilibrium to fulfill, each of which relates to a definite sort of transformation. Let us consider first that change which consists in the vaporization of the solvent. In accordance with our scheme of notation, the following conditions hold:

$$\nu_0 = -1,\ \nu_1 = 0,\ \nu_2 = 0,\ \cdots \nu_0' = \frac{m_0}{m_0'},\ \nu_1' = 0,\ \nu_2' = 0,\ \cdots,$$

and, therefore, the condition of equilibrium (6) becomes:

$$-\log c_0 + \frac{m_0}{m_0'} \log c_0' = \log K,$$

or, if one substitutes:

$$c_0 = 1 - \frac{n_1 + n_2 + \cdots}{n_0} \quad \text{and} \quad c_0' = 1 - \frac{n_1' + n_2' + \cdots}{n_0'},$$

$$\frac{n_1 + n_2 + \cdots}{n_0} - \frac{m_0}{m_0'} \cdot \frac{n_1' + n_2' + \cdots}{n_0'} = \log K.$$

If we treat this equation upon equation (9) as a model, there results an equation similar to (10):

$$T - T_0 = \left(\frac{n_1 + n_2 + \cdots}{n_0 m_0} - \frac{n_1' + n_2' + \cdots}{n_0' m_0'} \right) \frac{R T_0^2 m_0}{\Delta Q}.$$

Here ΔQ is the heat effect in the vaporization of one molecule of the solvent and, therefore, $\Delta Q/m_0$ is the heat effect in the vaporization of a unit mass of the solvent.

We remark, once more, that the solvent always occurs in the formula through the mass only, and not through the molecular number or the molecular weight, while, on the other hand, in the case of the dissolved substances, the molecular state is characteristic on account of their influence upon vaporization. Finally, the formula contains a generalization of the law of van't Hoff, stated above, for the raising of the boiling point, in that here in place of the number of dissolved molecules in the liquid, the difference between the number of dissolved molecules in unit mass of the liquid and in unit mass of the vapor appears. According as the unit mass of liquid or the unit mass of vapor contains more dissolved molecules, there results for the solution a raising or lowering of the boiling point; in the limiting case, when both quantities are equal, and the mixture therefore boils without changing, the change in boiling point becomes equal to zero. Of course, there are corresponding laws holding for the change in the vapor pressure.

Let us consider now a change which consists in the vaporization of a dissolved molecule. For this case we have in our notation

$$\nu_0 = 0,\ \nu_1 = -1,\ \nu_2 = 0 \cdots,\ \nu_0' = 0,\ \nu_1' = \frac{m_1}{m_1'},\ \nu_2' = 0,\ \cdots$$

and, in accordance with (6), for the condition of equilibrium:

$$-\log c_1 + \frac{m_1}{m_1'} \log c_1' = \log K$$

or:

$$\frac{c_1'^{\frac{m_1}{m_1'}}}{c_1} = K.$$

This equation expresses the Nernst law of distribution. If the dissolved substance possesses in both phases the same molecular weight ($m_1 = m_1'$), then, in a state of equilibrium a fixed ratio of the concentrations c_1 and c_1' in the liquid and in the vapor exists, which depends only upon the pressure and temperature. But, if the dissolved substance polymerises somewhat in the liquid, then the relation demanded in the last equation appears in place of the simple ratio.

VI. *The Dissolved Substance only Passes over into the Second Phase.*

This case is in a certain sense a special case of the one preceding. To it belongs that of the solubility of a slightly soluble salt, first investigated by van't Hoff, e. g., succinic acid in water. The symbol of this system is:

$$n_0 H_2O,\ n_1 H_6C_4O_4 \mid n_0' H_6C_4O_4,$$

in which we disregard the small dissociation of the acid solution. The concentrations of the individual molecular complexes are:

$$c_0 = \frac{n_0}{n_0 + n_1},\quad c_1 = \frac{n_1}{n_0 + n_1},\quad c_0' = \frac{n_0'}{n_0'} = 1.$$

For the precipitation of solid succinic acid we have:

$$\nu_0 = 0,\quad \nu_1 = -1,\quad \nu_0' = 1,$$

and, therefore, from the condition of equilibrium (6):

$$-\log c_1 = \log K,$$

hence, from (7):

$$\Delta Q = -RT^2 \frac{\partial \log c_1}{\partial T}.$$

By means of this equation van't Hoff calculated the heat of solution ΔQ from the solubility of succinic acid at 0° and at 8.5° C. The corresponding numbers were 2.88 and 4.22 in an arbitrary unit. Approximately, then:

$$\frac{\partial \log c_1}{\partial T} = \frac{\overset{e}{\log}\, 4.22 - \overset{e}{\log}\, 2.88}{8.5} = 0.04494,$$

from which for $T = 273$:

$$\Delta Q = -1.98 \cdot 273^2 \cdot 0.04494 = -6{,}600 \text{ cal.},$$

that is, in the precipitation of a molecule of succinic acid, 6,600 cal. are given out to the surroundings. Berthelot found, however, through direct measurement, 6,700 calories for the heat of solution.

The absorption of a gas also comes under this head, e. g. carbonic acid, in a liquid of relatively unnoticeable smaller vapor pressure, e. g., water at not too high a temperature. The symbol of the system is then

$$n_0 H_2O,\ n_1 CO_2 \mid n_0' CO_2.$$

The vaporization of a molecule CO_2 corresponds to the values

$$\nu_0 = 0, \quad \nu_1 = -1, \quad \nu_0' = 1.$$

The condition of equilibrium is therefore again:

$$-\log c_1 = \log K,$$

i. e., at a fixed temperature and a fixed pressure the concentration c_1 of the gas in the solution is constant. The change of the concentration with p and T is obtained through substitution in equation (7). It follows from this that:

$$\frac{\partial \log c_1}{\partial p} = \frac{\Delta V}{RT}, \quad \frac{\partial \log c_1}{\partial T} = -\frac{\Delta Q}{RT^2}.$$

ΔV is the change in volume of the system which occurs in the isobaric-isothermal vaporization of a molecule of CO_2, ΔQ the quantity of heat absorbed in the process from outside. Now, since ΔV represents approximately the volume of a molecule of gaseous carbonic acid, we may put approximately:

$$\Delta V = \frac{RT}{p},$$

and the equation gives:

$$\frac{\partial \log c_1}{\partial p} = \frac{1}{p},$$

which integrated, gives:

$$\log c_1 = \log p + \text{const.}, \quad c_1 = C \cdot p,$$

i. e., the concentration of the dissolved gas is proportional to the pressure of the free gas above the solution (law of Henry and Bunsen). The factor of proportionality C, which furnishes a measure of the solubility of the gas, depends upon the heat effect in quite the same manner as in the example previously considered.

A number of no less important relations are easily derived as by-products of those found above, e. g., the Nernst laws concerning the influence of solubility, the Arrhenius theory of isohydric solutions, etc. All such may be obtained through the application of the general condition of equilibrium (6). In conclusion, there is one other case that I desire to treat here. In the historical development of the theory this has played a particularly important rôle.

VII. *Osmotic Pressure.*

We consider now a dilute solution separated by a membrane (permeable with regard to the solvent but impermeable as regards the dissolved substance) from the pure solvent (in the same state of aggregation), and inquire as to the condition of equilibrium. The symbol of the system considered we may again take as

$$n_0 m_0,\ n_1 m_1,\ n_2 m_2,\ \cdots \mid n_0' m_0.$$

The condition of equilibrium is also here again expressed by equation (6), valid for a change of state in which the temperature and the pressure in each phase is maintained constant. The only difference with respect to the cases treated earlier is this, that here, in the presence of a separating membrane between two phases, the pressure p in the first phase may be different from the pressure p' in the second phase, whereby by "pressure," as always, is to be understood the ordinary hydrostatic or manometric pressure.

The proof of the applicability of equation (6) is found in the same way as this equation was derived above, proceeding from the principle of increase of entropy. One has but to remember that, in the somewhat more general case here considered, the external work in a given change is represented by the sum $pdV + p'dV'$, where V and V' denote the volumes of the two individual phases, while before V denoted the total volume of all phases. Accordingly, we use, instead of (7), to express the dependence of the constant K in (6) upon the pressure:

$$\frac{\partial \log K}{\partial p} = -\frac{\Delta V}{RT}, \quad \frac{\partial \log K}{\partial p'} = -\frac{\Delta V'}{RT}. \tag{11}$$

We have here to do with the following change:

$$\nu_0 = -1, \quad \nu_1 = 0, \quad \nu_2 = 0, \quad \cdots, \quad \nu_0' = 1,$$

whereby is expressed, that a molecule of the solvent passes out of the solution through the membrane into the pure solvent. Hence, in accordance with (6):

$$-\log c_0 = \log K,$$

or, since

$$c_0 = 1 - \frac{n_1 + n_2 + \cdots}{n_0}, \quad \frac{n_1 + n_2 + \cdots}{n_0} = \log K.$$

Here K depends only upon T, p and p'. If a pure solvent were present upon both sides of the membrane, we should have $c_0 = 1$, and $p = p'$; consequently:

$$(\log K)_{p=p'} = 0,$$

and by subtraction of the last two equations:

$$\frac{n_1 + n_2 + \cdots}{n_0} = \log K - (\log K)_{p=p'} = \frac{\partial \log K}{\partial p}(p - p')$$

and in accordance with (11):

$$\frac{n_1 + n_2 + \cdots}{n_0} = -(p - p') \cdot \frac{\Delta V}{RT}.$$

Here ΔV denotes the change in volume of the solution due to the loss of a molecule of the solvent ($\nu_0 = -1$). Approximately then:

$$-\Delta V \cdot n_0 = V,$$

the volume of the whole solution, and

$$\frac{n_1 + n_2 + \cdots}{n_0} = (p - p') \cdot \frac{V}{RT}.$$

If we call the difference $p - p'$, the osmotic pressure of the solution, this equation contains the well known law of osmotic pressure, due to van't Hoff.

The equations here derived, which easily permit of multiplication and generalization, have, of course, for the most part not been derived in the ways described above, but have been derived, either directly from experiment, or theoretically from the consideration of special reversible isothermal cycles to which the thermodynamic law was applied, that in such a cyclic process not only the algebraic sum of the work produced and the heat produced, but that also each of these two quantities separately, is equal to zero (first lecture, p. 19). The employment of a cyclic process has the advantage over the procedure here proposed, that in it the connection between the directly measurable quantities and the requirements of the laws of thermodynamics succinctly appears in each case; but for each individual case a satisfactory cyclic process must be imagined, and one has not always the certain assurance that the thermodynamic realization of the cyclic process also actually supplies all the conditions of equilibrium. Furthermore, in the process of calculation certain terms of considerable weight frequently appear as empty ballast, since they disappear at the end in the summation over the individual phases of the process.

On the other hand, the significance of the process here employed consists therein, that the necessary and sufficient conditions of equilibrium for each individually considered case appear collectively in the single equation (6), and that they are derived collectively from it in a direct manner through an unambiguous procedure. The more complicated the systems considered are, the more apparent becomes the advantage of this method, and there is no doubt in my mind that in chemical circles it will be more and more employed, especially, since in general it is now the custom to deal directly with the energies, and not with cyclic processes, in the calculation of heat effects in chemical changes.

THIRD LECTURE.

The Atomic Theory of Matter.

The problem with which we shall be occupied in the present lecture is that of a closer investigation of the atomic theory of matter. It is, however, not my intention to introduce this theory with nothing further, and to set it up as something apart and disconnected with other physical theories, but I intend above all to bring out the peculiar significance of the atomic theory as related to the present general system of theoretical physics; for in this way only will it be possible to regard the whole system as one containing within itself the essential compact unity, and thereby to realize the principal object of these lectures.

Consequently it is self evident that we must rely on that sort of treatment which we have recognized in last week's lecture as fundamental. That is, the division of all physical processes into reversible and irreversible processes. Furthermore, we shall be convinced that the accomplishment of this division is only possible through the atomic theory of matter, or, in other words, that irreversibility leads of necessity to atomistics.

I have already referred at the close of the first lecture to the fact that in pure thermodynamics, which knows nothing of an atomic structure and which regards all substances as absolutely continuous, the difference between reversible and irreversible processes can only be defined in one way, which a priori carries a provisional character and does not withstand penetrating analysis. This appears immediately evident when one reflects that the purely thermodynamic definition of irreversibility which proceeds from the impossibility of the realization of certain changes in nature, as, e. g., the transformation of heat into work without compensation, has at the outset assumed a definite limit to man's mental capacity, while, however, such a limit is not indicated in reality. On the contrary: mankind is making every endeavor to press beyond the present boundaries of its capacity, and we hope that later on many things will be attained which, perhaps, many regard at present as impossible of accomplishment. Can it not happen then that a process, which up to the present has been regarded as irreversible, may be proved, through a new discovery or invention, to be reversible? In this case the whole structure of the second law would undeniably collapse, for the irreversibility of a single process conditions that of all the others.

It is evident then that the only means to assure to the second law real meaning consists in this, that the idea of irreversibility be made independent of any relationship to man and especially of all technical relations.

Now the idea of irreversibility harks back to the idea of entropy; for a process is irreversible when it is connected with an increase of entropy. The problem is hereby referred back to a proper improvement of the definition of entropy. In accordance with the original definition of Clausius, the entropy is measured by means of a certain reversible process, and the weakness of this definition rests upon the fact that many such reversible processes, strictly speaking all, are not capable of being carried out in practice. With some reason it may be objected that we have here to do, not with an actual process and an actual physicist, but only with ideal processes, so-called thought experiments, and with an ideal physicist who operates with all the experimental methods with absolute accuracy. But at this point the difficulty is encountered: How far do the physicist's ideal measurements of this sort suffice? It may be understood, by passing to the limit, that a gas is compressed by a pressure which is equal to the pressure of the gas, and is heated by a heat reservoir which possesses the same temperature as the gas, but, for example, that a saturated vapor shall be transformed through isothermal compression in a reversible manner to a liquid without at any time a part of the vapor being condensed, as in certain thermodynamic considerations is supposed, must certainly appear doubtful. Still more striking, however, is the liberty as regards thought experiments, which in physical chemistry is granted the theorist. With his semi-permeable membranes, which in reality are only realizable under certain special conditions and then only with a certain approximation, he separates in a reversible manner, not only all possible varieties of molecules, whether or not they are in stable or unstable conditions, but he also separates the oppositely charged ions from one another and from the undissociated molecules, and he is disturbed, neither by the enormous electrostatic forces which resist such a separation, nor by the circumstance that in reality, from the beginning of the separation, the molecules become in part dissociated while the ions in part again combine. But such ideal processes are necessary throughout in order to make possible the comparison of the entropy of the undissociated molecules with the entropy of the dissociated molecules; for the law of thermodynamic equilibrium does not permit in general of derivation in any other way, in case one wishes to retain pure thermodynamics as a basis. It must be considered remarkable that all these ingenious thought processes have so well found confirmation of their results in experience, as is shown by the examples considered by us in the last lecture.

If now, on the other hand, one reflects that in all these results every reference to the possibility of actually carrying out each ideal process has disappeared—there are certainly left relations between directly measurable quantities only, such as temperature, heat effect, concentration, etc.—the presumption forces itself upon one that perhaps the introduction as above of such ideal processes is at bottom a round-about method, and that the peculiar import of the principle of increase of entropy with all its consequences can be evolved from the original idea of irreversibility or, just as well, from the impossibility of perpetual motion of the second kind, just as the principle of conservation of energy has been evolved from the law of impossibility of perpetual motion of the first kind.

This step: to have completed the emancipation of the entropy idea from the experimental art of man and the elevation of the second law thereby to a real principle, was the scientific life's work of Ludwig Boltzmann. Briefly stated, it consisted in general of referring back the idea of entropy to the idea of probability. Thereby is also explained, at the same time, the significance of the above (p. 17) auxiliary term used by me; "preference" of nature for a definite state. Nature prefers the more probable states to the less probable, because in nature processes take place in the direction of greater probability. Heat goes from a body at higher temperature to a body at lower temperature because the state of equal temperature distribution is more probable than a state of unequal temperature distribution.

Through this conception the second law of thermodynamics is removed at one stroke from its isolated position, the mystery concerning the preference of nature vanishes, and the entropy principle reduces to a well understood law of the calculus of probability.

The enormous fruitfulness of so "objective" a definition of entropy for all domains of physics I shall seek to demonstrate in the following lectures. But today we have principally to do with the proof of its admissibility; for on closer consideration we shall immediately perceive that the new conception of entropy at once introduces a great number of questions, new requirements and difficult problems. The first requirement is the introduction of the atomic hypothesis into the system of physics. For, if one wishes to speak of the probability of a physical state, i. e., if he wishes to introduce the probability for a given state as a definite quantity into the calculation, this can only be brought about, as in cases of all probability calculations, by referring the state back to a variety of possibilities; i. e., by considering a finite number of a priori equally likely configurations (complexions) through each of which the state considered may be realized. The greater the number of complexions, the greater is the probability of the state. Thus, e. g., the probability of throwing a total of four with two ordinary six-sided dice is found through counting the complexions by which the throw with a total of four may be realized. Of these there are three complexions:

with the first die, 1, with the second die, 3,
with the first die, 2, with the second die, 2,
with the first die, 3, with the second die, 1.

On the other hand, the throw of two is only realized through a single complexion. Therefore, the probability of throwing a total of four is three times as great as the probability of throwing a total of two.

Now, in connection with the physical state under consideration, in order to be able to differentiate completely from one another the complexions realizing it, and to associate it with a definite reckonable number, there is obviously no other means than to regard it as made up of numerous discrete homogeneous elements —for in perfectly continuous systems there exist no reckonable elements—and hereby the atomistic view is made a fundamental requirement. We have, therefore, to regard all bodies in nature, in so far as they possess an entropy, as constituted of atoms, and we therefore arrive in physics at the same conception of matter as that which obtained in chemistry for so long previously.

But we can immediately go a step further yet. The conclusions reached hold, not only for thermodynamics of material bodies, but also possess complete validity for the processes of heat radiation, which are thus referred back to the second law of thermodynamics. That radiant heat also possesses an entropy follows from the fact that a body which emits radiation into a surrounding diathermanous medium experiences a loss of heat and, therefore, a decrease of entropy. Since the total entropy of a physical system can only increase, it follows that one part of the entropy of the whole system, consisting of the body and the diathermanous medium, must be contained in the radiated heat. If the entropy of the radiant heat is to be referred back to the notion of probability, we are forced, in a similar way as above, to the conclusion that for radiant heat the atomic conception possesses a definite meaning. But, since radiant heat is not directly connected with matter, it follows that this atomistic conception relates, not to matter, but only to energy, and hence, that in heat radiation certain energy elements play an essential rôle. Even though this conclusion appears so singular and even though in many circles today vigorous objection is strongly urged against it, in the long run physical research will not be able to withhold its sanction from it, and the less, since it is confirmed by experience in quite a satisfactory manner. We shall return to this point in the lectures on heat radiation. I desire here only to mention that the novelty involved by the introduction of atomistic conceptions into the theory of heat radiation is by no means so revolutionary as, perhaps, might appear at the first glance. For there is, in my opinion at least, nothing which makes necessary the consideration of the heat processes in a complete vacuum as atomic, and it suffices to seek the atomistic features at the source of radiation, i. e., in those processes which have their play in the centres of emission and absorption of radiation. Then the Maxwellian electrodynamic differential equations can retain completely their validity for the vacuum, and, besides, the discrete elements of heat radiation are relegated exclusively to a domain which is still very mysterious and where there is still present plenty of room for all sorts of hypotheses.

Returning to more general considerations, the most important question comes up as to whether, with the introduction of atomistic conceptions and with the reference of entropy to probability, the content of the principle of increase of entropy is exhaustively comprehended, or whether still further physical hypotheses are required in order to secure the full import of that principle. If this important question had been settled at the time of the introduction of the atomic theory into thermodynamics, then the atomistic views would surely have been spared a large number of conceivable misunderstandings and justifiable attacks. For it turns out, in fact—and our further considerations will confirm this conclusion—that there has as yet nothing been done with atomistics which in itself requires much more than an essential generalization, in order to guarantee the validity of the second law.

We must first reflect that, in accordance with the central idea laid down in the first lecture (p. 7), the second law must possess validity as an objective physical law, independently of the individuality of the physicist. There is nothing to hinder us from imagining a physicist—we shall designate him a "microscopic" observer—whose senses are so sharpened that he is able to recognize each individual atom and to follow it in its motion. For this observer each atom moves exactly in accordance with the elementary laws which general dynamics lays down for it, and these laws allow, so far as we know, of an inverse performance of every process. Accordingly, here again the question is neither one of probability nor of entropy and its increase. Let us imagine, on the other hand, another observer, designated a "macroscopic" observer, who regards an ensemble of atoms as a homogeneous gas, say, and consequently applies the laws of thermodynamics to the mechanical and thermal processes within it. Then, for such an observer, in accordance with the second law, the process in general is an irreversible process. Would not now the first observer be justified in saying: "The reference of the entropy to probability has its origin in the fact that irreversible processes ought to be explained through reversible processes. At any rate, this procedure appears to me in the highest degree dubious. In any case, I declare each change of state which takes place in the ensemble of atoms designated a gas, as reversible, in opposition to the macroscopic observer." There is not the slightest thing, so far as I know, that one can urge against the validity of these statements. But do we not thereby place ourselves in the painful position of the judge who declared in a trial the correctness of the position of each separately of two contending parties and then, when a third contends that only one of the parties could emerge from the process victorious, was obliged to declare him also correct? Fortunately we find ourselves in a more favorable position. We can certainly mediate between the two parties without its being necessary for one or the other to give up his principal point of view. For closer consideration shows that the whole controversy rests upon a misunderstanding—a new proof of how necessary it is before one begins a controversy to come to an understanding with his opponent concerning the subject of the quarrel. Certainly, a given change of state cannot be both reversible and irreversible. But the one observer connects a wholly different idea with the phrase "change of state" than the other. What is then, in general, a change of state? The state of a physical system cannot well be otherwise defined than as the aggregate of all those physical quantities, through whose instantaneous values the time changes of the quantities, with given boundary conditions, are uniquely determined. If we inquire now, in accordance with the import of this definition, of the two observers as to what they understand by the state of the collection of atoms or the gas considered, they will give quite different answers. The microscopic observer will mention those quantities which determine the position and the velocities of all the individual atoms. There are present in the simplest case, namely, that in which the atoms may be considered as material points, six times as many quantities as atoms, namely, for each atom the three coordinates and the three velocity components, and in the case of combined molecules, still more quantities. For him the state and the progress of a process is then first determined when all these various quantities are individually given. We shall designate the state defined in this way the "micro-state." The macroscopic observer, on the other hand, requires fewer data. He will say that the state of the homogeneous gas considered by him is determined by the density, the visible velocity and the temperature at each point of the gas, and he will expect that, when these quantities are given, their time variations and, therefore, the progress of the process, to be completely determined in accordance with the two laws of thermo-dynamics, and therefore accompanied by an increase in entropy. In this connection he can call upon all the experience at his disposal, which will fully confirm his expectation. If we call this state the "macro-state," it is clear that the two laws: "the micro-changes of state are reversible" and "the macro-changes of state are irreversible," lie in wholly different domains and, at any rate, are not contradictory.

But now how can we succeed in bringing the two observers to an understanding? This is a question whose answer is obviously of fundamental significance for the atomic theory. First of all, it is easy to see that the macro-observer reckons only with mean values; for what he calls density, visible velocity and temperature of the gas are, for the micro-observer, certain mean values, statistical data, which are derived from the space distribution and from the velocities of the atoms in an appropriate manner. But the micro-observer cannot operate with these mean values alone, for, if these are given at one instant of time, the progress of the process is not determined throughout; on the contrary: he can easily find with given mean values an enormously large number of individual values for the positions and the velocities of the atoms, all of which correspond with the same mean values and which, in spite of this, lead to quite different processes with regard to the mean values. It follows from this of necessity that the micro-observer must either give up the attempt to undertand the unique progress, in accordance with experience, of the macroscopic changes of state—and this would be the end of the atomic theory—or that he, through the introduction of a special physical hypothesis, restrict in a suitable manner the manifold of micro-states considered by him. There is certainly nothing to prevent him from assuming that not all conceivable micro-states are realizable in nature, and that certain of them are in fact thinkable, but never actually realized. In the formularization of such a hypothesis, there is of course no point of departure to be found from the principles of dynamics alone; for pure dynamics leaves this case undetermined. But on just this account any dynamical hypothesis, which involves nothing further than a closer specification of the micro-states realized in nature, is certainly permissible. Which hypothesis is to be given the preference can only be decided through comparison of the results to which the different possible hypotheses lead in the course of experience.

In order to limit the investigation in this way, we must obviously fix our attention only upon all imaginable configurations and velocities of the individual atoms which are compatible with determinate values of the density, the velocity and the temperature of the gas, or in other words: we must consider all the micro-states which belong to a determinate macro-state, and must investigate the various kinds of processes which follow in accordance with the fixed laws of dynamics from the different micro-states. Now, precise calculation has in every case always led to the important result that an enormously large number of these different micro-processes relate to one and the same macro-process, and that only proportionately few of the same, which are distinguished by quite special exceptional conditions concerning the positions and the velocities of neighboring atoms, furnish exceptions. Furthermore, it has also shown that one of the resulting macro-processes is that which the macroscopic observer recognizes, so that it is compatible with the second law of thermodynamics.

Here, manifestly, the bridge of understanding is supplied. The micro-observer needs only to assimilate in his theory the physical hypothesis that all those special cases in which special exceptional conditions exist among the neighboring configurations of interacting atoms do not occur in nature, or, in other words, that the micro-states are in elementary disorder. Then the uniqueness of the macroscopic process is assured and with it, also, the fulfillment of the principle of increase of entropy in all directions.

Therefore, it is not the atomic distribution, but rather the hypothesis of elementary disorder, which forms the real kernel of the principle of increase of entropy and, therefore, the preliminary condition for the existence of entropy. Without elementary disorder there is neither entropy nor irreversible process.[1] Therefore, a single atom can never possess an entropy; for we cannot speak of disorder in connection with it. But with a fairly large number of atoms, say 100 or 1,000, the matter is quite different. Here, one can certainly speak of a disorder, in case that the values of the coordinates and the velocity components are distributed among the atoms in accordance with the laws of accident. Then it is possible to calculate the probability for a given state. But how is it with regard to the increase of entropy? May we assert that the motion of 100 atoms is irreversible? Certainly not; but this is only because the state of 100 atoms cannot be defined in a thermodynamic sense, since the process does not proceed in a unique manner from the standpoint of a macro-observer, and this requirement forms, as we have seen above, the foundation and preliminary condition for the definition of a thermodynamic state.

If one therefore asks: How many atoms are at least necessary in order that a process may be considered irreversible?, the answer is: so many atoms that one may form from them definite mean values which define the state in a macroscopic sense. One must reflect that to secure the validity of the principle of increase of entropy there must be added to the condition of elementary disorder still another, namely, that the number of the elements under consideration be sufficiently large to render possible the formation of definite mean values. The second law has a meaning for these mean values only; but for them, it is quite

[1] To those physicists who, in spite of all this, regard the hypothesis of elementary disorder as gratuitous or as incorrect, I wish to refer the simple fact that in every calculation of a coefficient of friction, of diffusion, or of heat conduction, from molecular considerations, the notion of elementary disorder is employed, whether tacitly or otherwise, and that it is therefore essentially more correct to stipulate this condition instead of ignoring or concealing it. But he who regards the hypothesis of elementary disorder as self-evident, should be reminded that, in accordance with a law of H. Poincaré, the precise investigation concerning the foundation of which would here lead us too far, the assumption of this hypothesis for all times is unwarranted for a closed space with absolutely smooth walls,—an important conclusion, against which can only be urged the fact that absolutely smooth walls do not exist in nature.

exact, just as exact as the law of the calculus of probability, that the mean value, so far as it may be defined, of a sufficiently large number of throws with a six-sided die, is $3\frac{1}{2}$.

These considerations are, at the same time, capable of throwing light upon questions such as the following: Does the principle of increase of entropy possess a meaning for the so-called Brownian molecular movement of a suspended particle? Does the kinetic energy of this motion represent useful work or not? The entropy principle is just as little valid for a single suspended particle as for an atom, and therefore is not valid for a few of them, but only when there is so large a number that definite mean values can be formed. That one is able to see the particles and not the atoms makes no material difference; because the progress of a process does not depend upon the power of an observing instrument. The question with regard to useful work plays no rôle in this connection; strictly speaking, this possesses, in general, no objective physical meaning. For it does not admit of an answer without reference to the scheme of the physicist or technician who proposes to make use of the work in question. The second law, therefore, has fundamentally nothing to do with the idea of useful work (cf. first lecture, p. 15).

But, if the entropy principle is to hold, a further assumption is necessary, concerning the various disordered elements,—an assumption which tacitly is commonly made and which we have not previously definitely expressed. It is, however, not less important than those referred to above. The elements must actually be of the same kind, or they must at least form a number of groups of like kind, e. g., constitute a mixture in which each kind of element occurs in large numbers. For only through the similarity of the elements does it come about that order and law can result in the larger from the smaller. If the molecules of a gas be all different from one another, the properties of a gas can never show so simple a law-abiding behavior as that which is indicated by thermodynamics. In fact, the calculation of the probability of a state presupposes that all complexions which correspond to the state are a priori equally likely. Without this condition one is just as little able to calculate the probability of a given state as, for instance, the probability of a given throw with dice whose sides are unequal in size. In summing up we may therefore say: the second law of thermodynamics in its objective physical conception, freed from anthropomorphism, relates to certain mean values which are formed from a large number of disordered elements of the same kind.

The validity of the principle of increase of entropy and of the irreversible progress of thermodynamic processes in nature is completely assured in this formularization. After the introduction of the hypothesis of elementary disorder, the microscopic observer can no longer confidently assert that each process considered by him in a collection of atoms is reversible; for the motion occurring in the reverse order will not always obey the requirements of that hypothesis. In fact, the motions of single atoms are always reversible, and thus far one may say, as before, that the irreversible processes appear reduced to a reversible process, but the phenomenon as a whole is nevertheless irreversible, because upon reversal the disorder of the numerous individual elementary processes would be eliminated. Irreversibility is inherent, not in the individual elementary processes themselves, but solely in their irregular constitution. It is this only which guarantees the unique change of the macroscopic mean values.

Thus, for example, the reverse progress of a frictional process is impossible, in that it would presuppose elementary arrangement of interacting neighboring molecules. For the collisions between any two molecules must thereby possess a certain distinguishing character, in that the velocities of two colliding molecules depend in a definite way upon the place at which they meet. In this way only can it happen that in collisions like directed velocities ensue and, therefore, visible motion.

Previously we have only referred to the principle of elementary disorder in its application to the atomic theory of matter. But it may also be assumed as valid, as I wish to indicate at this point, on quite the same grounds as those holding in the case of matter, for the theory of radiant heat. Let us consider, e. g., two bodies at different temperatures between which exchange of heat occurs through radiation. We can in this case also imagine a microscopic observer, as opposed to the ordinary macroscopic observer, who possesses insight into all the particulars of electromagnetic processes which are connected with emission and absorption, and the propagation of heat rays. The microscopic observer would declare the whole process reversible because all electrodynamic processes can also take place in the reverse direction, and the contradiction may here be referred back to a difference in definition of the state of a heat ray. Thus, while the macroscopic observer completely defines a monochromatic ray through direction, state of polarization, color, and intensity, the microscopic observer, in order to possess a complete knowledge of an electromagnetic state, necessarily requires the specification of all the numerous irregular variations of amplitude and phase to which the most homogeneous heat ray is actually subject. That such irregular variations actually exist follows immediately from the well known fact that two rays of the same color never interfere, except when they originate in the same source of light. But until these fluctuations are given in all particulars, the micro-observer can say nothing with regard to the progress of the process. He is also unable to specify whether the exchange of heat radiation between the two bodies leads to a decrease or to an increase of their difference in temperature. The principle of elementary disorder first furnishes the adequate criterion of the tendency of the radiation process, i. e., the warming of the colder body at the expense of the warmer, just as the same principle conditions the irreversibility of exchange of heat through conduction. However, in the two cases compared, there is indicated an essential difference in the kind of the disorder. While in heat conduction the disordered elements may be represented as associated with the various molecules, in heat radiation there are the numerous vibration periods, connected with a heat ray, among which the energy of radiation is irregularly distributed. In other words: the disorder among the molecules is a material one, while in heat radiation it is one of energy distribution. This is the most important difference between the two kinds of disorder; a common feature exists as regards the great number of uncoordinated elements required. Just as the entropy of a body is defined as a function of the macroscopic state, only when the body contains so many atoms that from them definite mean values may be formed, so the entropy principle only possesses a meaning with regard to a heat ray when the ray comprehends so many periodic vibrations, i. e., persists for so long a time, that a definite mean value for the intensity of the ray may be obtained from the successive irregular fluctuating amplitudes.

Now, after the principle of elementary disorder has been introduced and accepted by us as valid throughout nature, the fundamental question arises as to the calculation of the probability of a given state, and the actual derivation of the entropy therefrom. From the entropy all the laws of thermodynamic states of equilibrium, for material substances, and also for energy radiation, may be uniquely derived. With regard to the connection between entropy and probability, this is inferred very simply from the law that the probability of two independent configurations is represented by the product of the individual probabilities:

$$W = W_1 \cdot W_2,$$

while the entropy S is represented by the sum of the individual entropies:

$$S = S_1 + S_2.$$

Accordingly, the entropy is proportional to the logarithm of the probability:

$$S = k \log W. \tag{12}$$

k is a universal constant. In particular, it is the same for atomic as for radiation configurations, for there is nothing to prevent us assuming that the configuration designated by 1 is atomic, while that designated by 2 is a radiation configuration. If k has been calculated, say with the aid of radiation measurements, then k must have the same value for atomic processes. Later we shall follow this procedure, in order to utilize the laws of heat radiation in the kinetic theory of gases. Now, there remains, as the last and most difficult part of the problem, the calculation of the probability W of a given physical configuration in a given macroscopic state. We shall treat today, by way of preparation for the quite general problem to follow, the simple problem: to specify the probability of a given state for a single moving material point, subject to given conservative forces. Since the state depends upon 6 variables: the 3 generalized coordinates φ_1, φ_2, φ_3, and the three corresponding velocity components $\dot{\varphi}_1$, $\dot{\varphi}_2$, $\dot{\varphi}_3$, and since all possible values of these 6 variables constitute a continuous manifold, the probability sought is, that these 6 quantities shall lie respectively within certain infinitely small intervals, or, if one thinks of these 6 quantities as the rectilinear orthogonal coordinates of a point in an ideal six-dimensional space, that this ideal "state point" shall fall within a given, infinitely small "state domain." Since the domain is infinitely small, the probability will be proportional to the magnitude of the domain and therefore proportional to

$$\int d\varphi_1 \cdot d\varphi_2 \cdot d\varphi_3 \cdot d\dot{\varphi}_1 \cdot d\dot{\varphi}_2 \cdot d\dot{\varphi}_3.$$

But this expression cannot serve as an absolute measure of the probability, because in general it changes in magnitude with the time, if each state point moves in accordance with the laws of motion of material points, while the probability of a state which follows of necessity from another must be the same for the one as the other. Now, as is well known, another integral quite similarly formed, may be specified in place of the one above, which possesses the special property of not changing in value with the time. It is only necessary to employ, in addition to the general coordinates φ_1, φ_2, φ_3, the three so-called momenta

$\psi_1, \psi_2. \psi_3$, in place of the three velocities $\dot{\varphi}_1, \dot{\varphi}_2, \dot{\varphi}_3$, as the determining coordinates of the state. These are defined in the following way:

$$\psi_1 = \left(\frac{\partial H}{\partial \dot{\varphi}_1}\right)_\phi, \quad \psi_2 = \left(\frac{\partial H}{\partial \dot{\varphi}_2}\right)_\phi, \quad \dot{\varphi}_3 = \left(\frac{\partial H}{\partial \dot{\varphi}_3}\right)_\phi,$$

wherein H denotes the kinetic potential (Helmholz). Then, in Hamiltonian form, the equations of motion are:

$$\dot{\psi}_1 = \frac{d\psi_1}{dt} = -\left(\frac{\partial E}{\partial \varphi_1}\right)_\psi, \quad \cdots, \quad \dot{\varphi}_1 = \frac{d\varphi_1}{dt} = \left(\frac{\partial E}{\partial \psi_1}\right)_\phi, \quad \cdots,$$

(E is the energy), and from these equations follows the "condition of incompressibility":

$$\frac{\partial \dot{\varphi}_1}{\partial \varphi_1} + \frac{\partial \dot{\psi}_1}{\partial \psi_1} + \cdots = 0.$$

Referring to the six-dimensional space represented by the coordinates $\varphi_1, \varphi_2, \varphi_3, \psi_1, \psi_2, \psi_3$, this equation states that the magnitude of an arbitrarily chosen state domain, viz.:

$$\int d\varphi_1 \cdot d\varphi_2 \cdot d\varphi_3 \cdot d\psi_1 \cdot d\psi_2 \cdot d\psi_3$$

does not change with the time, when each point of the domain changes its position in accordance with the laws of motion of material points. Accordingly, it is made possible to take the magnitude of this domain as a direct measure for the probability that the state point falls within the domain.

From the last expression, which can be easily generalized for the case of an arbitrary number of variables, we shall calulate later the probability of a thermodynamic state, for the case of radiant energy as well as that for material substances.

FOURTH LECTURE.

The Equation of State for a Monatomic Gas.

My problem today is to utilize the general fundamental laws concerning the concept of irreversibility, which we established in the lecture of yesterday, in the solution of a definite problem: the calculation of the entropy of an ideal monatomic gas in a given state, and the derivation of all its thermodynamic properties. The way in which we have to proceed is prescribed for us by the general definition of entropy:

$$S = k \log W. \tag{13}$$

The chief part of our problem is the calculation of W for a given state of the gas, and in this connection there is first required a more precise investigation of that which is to be understood as the state of the gas. Obviously, the state is to be taken here solely in the sense of the conception which we have called macroscopic in the last lecture. Otherwise, a state would possess neither probability nor entropy. Furthermore, we are not allowed to assume a condition of equilibrium for the gas. For this is characterized through the further special condition that the entropy for it is a maximum. Thus, an unequal distribution of density may exist in the gas; also, there may be present an arbitrary number of different currents, and in general no kind of equality between the various velocities of the molecules is to be assumed. The velocities, as the coordinates of the molecules, are rather to be taken a priori as quite arbitrarily given, but in order that the state, considered in a macroscopic sense, may be assumed as known, certain mean values of the densities and the velocities must exist. Through these mean values the state from a macroscopic standpoint is completely characterized.

The conditions mentioned will all be fulfilled if we consider the state as given in such manner that the number of molecules in a sufficiently small macroscopic space, but which, however, contains a very large number of molecules, is given, and furthermore, that the (likewise great) number of these molecules is given, which are found in a certain macroscopically small velocity domain, i. e., whose velocities lie within certain small intervals. If we call the coordinates x, y, z, and the velocity components $\dot{x}$, $\dot{y}$, $\dot{z}$, then this number will be proportional to[1]

$$dx \cdot dy \cdot dz \cdot d\dot{x} \cdot d\dot{y} \cdot d\dot{z} = \sigma.$$

It will depend, besides, upon a finite factor of proportionality which may be an arbitrarily given function $f(x, y, z, \dot{x}, \dot{y}, \dot{z})$ of the coordinates and the velocities, and which has only the one condition to fulfill that

$$\Sigma f \cdot \sigma = N, \tag{14}$$

where N denotes the total number of molecules in the gas. We are now concerned with the calculation of the probability W of that state of the gas which corresponds to the arbitrarily given distribution function f.

The probability that a given molecule possesses such coordinates and such velocities that it lies within the domain σ is expressed, in accordance with the final result of the previous lecture, by the magnitude of the corresponding elementary domain:

$$d\varphi_1 \cdot d\varphi_2 \cdot d\varphi_3 \cdot d\psi_1 \cdot d\psi_2 \cdot d\psi_3,$$

therefore, since here

$$\varphi_1 = x, \quad \varphi_2 = y, \quad \varphi_3 = z, \quad \psi_1 = m\dot{x}, \quad \psi_2 = m\dot{y}, \quad \psi_3 = m\dot{z},$$

[1] We can call σ a "macro-differential" in contradistinction to the micro-differentials which are infinitely small with reference to the dimensions of a molecule. I prefer this terminology for the discrimination between "physical" and "mathematical" differentials in spite of the inelegance of phrasing, because the macro-differential is also just as much mathematical as physical and the micro-differential just as much physical as mathematical.

(m the mass of a molecule) by

$$m^3\sigma.$$

Now we divide the whole of the six dimensional "state domain" containing all the molecules into suitable equal elementary domains of the magnitude $m^3\sigma$. Then the probability that a given molecule fall in a given elementary domain is equally great for all such domains. Let P denote the number of these equal elementary domains. Next, let us imagine as many dice as there are molecules present, i. e., N, and each die to be provided with P equal sides. Upon these P sides we imagine numbers 1, 2, 3, $\cdots$ to P, so that each of the P sides indicates a given elementary domain. Then each throw with the N dice corresponds to a given state of the gas, while the number of dice which show a given number corresponds to the molecules which lie in the elementary domain considered. In accordance with this, each single die can indicate with the same probability each of the numbers from 1 to P, corresponding to the circumstance that each molecule may fall with equal probability in any one of the P elementary domains. The probability W sought, of the given state of the molecules, corresponds, therefore, to the number of different kinds of throws (complexions) through which is realized the given distribution f. Let us take, e. g., N equal to 10 molecules (dice) and $P = 6$ elementary domains (sides) and let us imagine the state so given that there are

3 molecules in 1st elementary domain
4 molecules in 2d elementary domain
0 molecules in 3d elementary domain
1 molecule in 4th elementary domain
0 molecules in 5th elementary domain
2 molecules in 6th elementary domain,

then this state, e. g., may be realized through a throw for which the 10 dice indicate the following numbers:

1st	2d	3d	4th	5th	6th	7th	8th	9th	10th	
2	6	2	1	1	2	6	2	1	4.	(15)

Under each of the characters representing the ten dice stands the number which the die indicates in the throw. In fact,

3 dice show the figure 1
4 dice show the figure 2
0 dice show the figure 3
1 die shows the figure 4
0 dice show the figure 5
2 dice show the figure 6.

The state in question may likewise be realized through many other complexions of this kind. The number sought of all possible complexions is now found through consideration of the number series indicated in (15). For, since the number of molecules (dice) is given, the number series contains a fixed number of elements ($10 = N$). Furthermore, since the number of molecules falling in an elementary domain is given, each number, in all permissible complexions, appears equally often in the series. Finally, each change of the number configuration conditions a new complexion. The number of possible complexions or the probability W of the given state is therefore equal to the number of possible permutations with repetition under the conditions mentioned. In the simple example chosen, in accordance with a well known formula, the probability is

$$\frac{10!}{3!\,4!\,0!\,1!\,0!\,2!} = 12{,}600.$$

Therefore, in the general case:

$$W = \frac{N!}{\Pi(f \cdot \sigma)!}.$$

The sign Π denotes the product extended over all of the P elementary domains.

From this there results, in accordance with equation (13), for the entropy of the gas in the given state:

$$S = k \log N! - k\Sigma \log (f \cdot \sigma)!.$$

The summation is to be extended over all domains σ. Since $f \cdot \sigma$ is a large quantity, Stirling's formula may be employed for its factorial, which for a large number n is expressed by:

$$n! = \left(\frac{n}{e}\right)^n \sqrt{2\pi n}, \tag{16}$$

therefore, neglecting unimportant terms:

$$\log n! = n(\log n - 1);$$

and hence:

$$S = k \log N! - k\Sigma f\sigma(\log [f \cdot \sigma] - 1),$$

or, if we note that σ and $N = \Sigma f\sigma$ remain constant in all changes of state:

$$S = \text{const} - k\Sigma f \cdot \log f \cdot \sigma. \tag{17}$$

This quantity is, to the universal factor $(-k)$, the same as that which L. Boltzmann denoted by H, and which he showed to vary in one direction only for all changes of state.

In particular, we will now determine the entropy of a gas in a state of equilibrium, and inquire first as to that form of the law of distribution which corresponds to thermodynamic equilibrium. In accordance with the second law of thermodynamics, a state of equilibrium is characterized by the condition that with given values of the total volume V and the total energy E, the entropy S assumes its maximum value. If we assume the total volume of the gas

$$V = \int dx \cdot dy \cdot dz,$$

and the total energy

$$E = \frac{m}{2}\Sigma(\dot{x}^2 + \dot{y}^2 + \dot{z}^2)f\sigma \tag{18}$$

as given, then the condition:

$$\delta S = 0$$

must hold for the state of equilibrium, or, in accordance with (17):

$$\Sigma(\log f + 1) \cdot \delta f \cdot \sigma = 0, \tag{19}$$

wherein the variation δf refers to an arbitrary change in the law of distribution, compatible with the given values of N, V and E.

Now we have, on account of the constancy of the total number of molecules N, in accordance with (14):

$$\Sigma \delta f \cdot \sigma = 0$$

and, on account of the constancy of the total energy, in accordance with (18):

$$\Sigma(\dot{x}^2 + \dot{y}^2 + \dot{z}^2) \cdot \delta f \cdot \sigma = 0.$$

Consequently, for the fulfillment of condition (19) for all permissible values of δf, it is sufficient and necessary that

$$\log f + \beta(\dot{x}^2 + \dot{y}^2 + \dot{z}^2) = \text{const},$$

or:

$$f = \alpha e^{-\beta(\dot{x}^2+\dot{y}^2+\dot{z}^2)},$$

wherein α and β are constants. In the state of equilibrium, therefore, the space distribution of molecules is uniform, i. e., independent of x, y, z, and the distribution of velocities is the well known Maxwellian distribution.

The values of the constants α and β are to be found from those of N, V and E. For the substitution of the value found for f in (14) leads to:

$$N = V\alpha\left(\frac{\pi}{\beta}\right)^{\frac{3}{2}},$$

and the substitution of f in (18) leads to:

$$E = \tfrac{3}{4}Vm\frac{\alpha}{\beta}\left(\frac{\pi}{\beta}\right)^{\frac{3}{2}}.$$

From these equations it follows that:

$$\alpha = \frac{N}{V} \cdot \left(\frac{3mN}{4\pi E}\right)^{\frac{3}{2}}, \quad \beta = \frac{3mN}{4E},$$

and hence finally, in accordance with (17), the expression for the entropy S of the gas in a state of equilibrium with given values for N, V and E is:

$$S = \text{const} + kN(\tfrac{3}{2}\log E + \log V). \qquad (20)$$

The additive constant contains terms in N and m, but not in E and V.

The determination of the entropy here carried out permits now the specification directly of the complete thermodynamic behavior of the gas, viz., of the equation of state, and of the values of the specific heats. From the general thermodynamic definition of entropy:

$$dS = \frac{dE + pdV}{T}$$

are obtained the partial differential quotients of S with regard to E and V respectively:

$$\left(\frac{dS}{\partial E}\right)_V = \frac{1}{T}, \quad \left(\frac{\partial S}{\partial V}\right)_E = \frac{p}{T}.$$

Consequently, with the aid of (20):

$$\left(\frac{\partial S}{dE}\right)_V = \frac{3}{2}\frac{kN}{E} = \frac{1}{T}, \qquad (21)$$

and

$$\left(\frac{\partial S}{\partial V}\right)_E = \frac{kN}{V} = \frac{p}{T}. \qquad (22)$$

The second of these equations:

$$p = \frac{kNT}{V}$$

contains the laws of Boyle, Gay Lussac and Avogadro, the latter because the pressure depends only upon the number N, and not upon the constitution of the molecules. Writing it in the ordinary form:

$$p = \frac{RnT}{V},$$

where n denotes the number of gram molecules or mols of the gas, referred to $O_2 = 32g$, and R the absolute gas constant:

$$R = 8.315 \cdot 10^7 \frac{\text{erg}}{\text{deg}},$$

we obtain by comparison:

$$k = \frac{Rn}{N}. \tag{23}$$

If we denote the ratio of the mol number to the molecular number by ω, or, what is the same thing, the ratio of the molecular mass to the mol mass:

$$\omega = \frac{n}{N},$$

and hence:

$$k = \omega R. \tag{24}$$

From this, if ω is given, we can calculate the universal constant k, and conversely.

The equation (21) gives:

$$E = \tfrac{3}{2} kNT. \tag{25}$$

Now since the energy of an ideal gas is given by:

$$E = Anc_v T,$$

wherein c_v denotes in calories the heat capacity at constant volume of a mol, A the mechanical equivalent of heat:

$$A = 4.19 \cdot 10^7 \frac{\text{erg}}{\text{cal}},$$

it follows that:

$$c_v = \frac{3kN}{2An},$$

and, having regard to (23), we obtain:

$$c_v = \frac{3}{2}\frac{R}{A} = 3.0, \tag{26}$$

the mol heat in calories of any monatomic gas at constant volume.

For the mol heat c_p at constant pressure we have from the first law of thermodynamics

$$c_p - c_v = \frac{R}{A},$$

and, therefore, having regard to (26):

$$c_p = 5, \quad \frac{c_p}{c_v} = \tfrac{5}{3},$$

a known result for monatomic gases.

The mean kinetic energy L of a molecule is obtained from (25):

$$L = \frac{E}{N} = \tfrac{3}{2}kT. \tag{27}$$

You notice that we have derived all these relations through the identification of the mechanical with the thermodynamic expression for the entropy, and from this you recognize the fruitfulness of the method here proposed.

But a method can first demonstrate fully its usefulness when we utilize it, not only to derive laws which are already known, but when we apply it in domains for whose investigation there at present exist no other methods. In this connection its application affords various possibilities. Take the case of a monatomic gas which is not sufficiently attenuated to have the properties of the ideal state; there are here, as pointed out by J. D. van der Waals, two things to consider: (1) the finite size of the atoms, (2) the forces which act among the atoms. Taking account of these involves a change in the value of the probability and in the energy of the gas as well, and, so far as can now be shown, the corresponding change in the conditions for thermodynamic equilibrium leads to an equation of state which agrees with that of van der Waals. Certainly there is here a rich field for further investigations, of greater promise when experimental tests of the equation of state exist in larger number.

Another important application of the theory has to do with heat radiation, with which we shall be occupied the coming week. We shall proceed then in a similar way as here, and shall be able from the expression for the entropy of radiation to derive the thermodynamic properties of radiant heat.

Today we will refer briefly to the treatment of polyatomic gases. I have previously, upon good grounds, limited the treatment to monatomic molecules; for up to the present real difficulties appear to stand in the way of a generalization, from the principles employed by us, to include polyatomic molecules; in fact, if we wish to be quite frank, we must say that a satisfactory mechanical theory of polyatomic gases has not yet been found. Consequently, at present we do not know to what place in the system of theoretical physics to assign the processes within a molecule—the intra-molecular processes. We are obviously confronted by puzzling problems. A noteworthy and much discussed beginning was, it is true, made by Boltzmann, who introduced the most plausible assumption that for intra-molecular processes simple laws of the same kind hold as for the motion of the molecules themselves, *i. e.*, the general equations of dynamics. It is easy then, in fact, to proceed to the proof that for a monatomic gas the molecular heat c_v must be greater than 3 and that consequently, since the difference $c_p - c_v$ is always equal to 2, the ratio is

$$\frac{c_p}{c_v} = \frac{c_v + 2}{c_v} < \tfrac{5}{3}.$$

This conclusion is completely confirmed by experience. But this in itself does not confirm the assumption of Boltzmann; for, indeed, the same conclusion is reached very simply from the assumption that there exists intra-molecular energy which increases with the temperature. For then the molecular heat of a polyatomic gas must be greater by a corresponding amount than that of a monatomic gas.

Nevertheless, up to this point the Boltzmann theory never leads to contradiction with experience. But so soon as one seeks to draw special conclusions concerning the magnitude of the specific heats hazardous difficulties arise; I will refer to only one of them. If one assumes the Hamiltonian equations of mechanics as applicable to intra-molecular motions, he arrives of necessity at the law of "uniform distribution of energy," which asserts that under certain conditions, not essential to consider here, in a thermodynamic state of equilibrium the total energy of the gas is distributed uniformly among all the individual energy phases corresponding to the independent variables of state, or, as one may briefly say; the same amount of energy is associated with every independent variable of state. Accordingly, the mean energy of motion of the molecules $\frac{1}{2}kT$, corresponding to a given direction in space, is the same as for any other direction, and, moreover, the same for all the different kinds of molecules, and ions; also for all suspended particles (dust) in the gas, of whatever size, and, furthermore, the same for all kinds of motions of the constituents of a molecule relative to its centroid. If one now reflects that a molecule commonly contains, so far as we know, quite a large number of different freely moving constituents, certainly, that a normal molecule of a monatomic gas, e. g., mercury, possesses numerous freely moving electrons, then, in accordance with the law of uniform energy distribution, the intra-molecular energy must constitute a much larger fraction of the whole specific heat of the gas, and therefore c_p/c_v must turn out much smaller, than is consistent with the measured values. Thus, e. g., for an atom of mercury, in accordance with the measured value of $c_p/c_v = 5/3$, no part whatever oft he heat added may be assigned to the intra-molecular energy. Boltzmann and others, in order to eliminate this contradiction, have fixed upon the possibility that, within the time of observation of the specific heats, the vibrations of the constituents (of a molecule) do not change appreciably with respect to one another, and come later with their progressive motion so slowly into heat equilibrium that this process is no longer capable of detection through observation. Up to now no such delay in the establishment of a state of equilibrium has been observed. Perhaps it would be productive of results if in delicate measurements special attention were paid the question as to whether observations which take a longer time lead to a greater value of the mol-heat, or, what comes to the same thing, a smaller value of c_p/c_v, than observations lasting a shorter time.

If one has been made mistrustful through these considerations concerning the applicability of the law of uniform energy distribution to intra-molecular processes, the mistrust is accentuated upon the inclusion of the laws of heat radiation. I shall make mention of this in a later lecture.

When we pass from stable atoms to the unstable atoms of radioactive substances, the principles following from the kinetic gas theory lose their validity completely. For the striking failure of all attempts to find any influence of temperature upon radioactive phenomena shows us that an application here of the law of uniform energy distribution is certainly not warranted. It will, therefore, be safest meanwhile to offer no definite conjectures with regard to the nature and the laws of these noteworthy phenomena, and to leave this field for further development to experimental research alone, which, I may say, with every day throws new light upon the subject.

FIFTH LECTURE.

HEAT RADIATION. ELECTRODYNAMIC THEORY.

Last week I endeavored to point out that we find in the atomic theory a complete explanation for the whole content of the two laws of thermodynamics, if we, with Boltzmann, define the entropy by the probability, and I have further shown, in the example of an ideal monatomic gas, how the calculation of the probability, without any additional special hypothesis, enables us not only to find the properties of gases known from thermodynamics, but also to reach conclusions which lie essentially beyond those of pure thermodynamics. Thus, e. g., the law of Avogadro in pure thermodynamics is only a definition, while in the kinetic theory it is a necessary consequence; furthermore, the value of c_v, the mol-heat of a gas, is completely undetermined by pure thermodynamics, but from the kinetic theory it is of equal magnitude for all monatomic gases and, in fact, equal to 3, corresponding to our experimental knowledge. Today and tomorrow we shall be occupied with the application of the theory to radiant heat, and it will appear that we reach in this apparently quite isolated domain conclusions which a thorough test shows are compatible with experience. Naturally, we take as a basis the electro-magnetic theory of heat radiation, which regards the rays as electromagnetic waves of the same kind as light rays.

We shall utilize the time today in developing in bold outline the important consequences which follow from the electromagnetic theory for the characteristic quantities of heat radiation, and tomorrow seek to answer, through the calculation of the entropy, the question concerning the dependence of these quantities upon the temperature, as was done last week for ideal gases. Above all, we are concerned here with the determination of those quantities which at any place in a medium traversed by heat rays determine the state of the radiant heat. The state of radiation at a given place will not be represented by a vector which is determined by three components; for the energy flowing in a given direction is quite independent of that flowing in any other direction. In order to know the state of radiation, we must be able to specify, moreover, the energy which in the time dt flows through a surface element $d\sigma$ for every direction in space. This will be proportional to the magnitude of $d\sigma$, to the time dt, and to the cosine of the angle ϑ which the direction considered makes with the normal to $d\sigma$. But the quantity to be multiplied by $d\sigma \cdot dt \cdot \cos\vartheta$ will not be a finite quantity; for since the radiation through any point of $d\sigma$ passes in all directions, therefore the quantity will also depend upon the magnitude of the solid angle $d\Omega$, which we shall assume as the same for all points of $d\sigma$. In this manner we obtain for the energy which in the time dt flows through the surface element $d\sigma$ in the direction of the elementary cone $d\Omega$, the expression:

$$K d\sigma dt \cdot \cos\vartheta \cdot d\Omega. \tag{28}$$

K is a positive function of place, of time and of direction, and is for unpolarized light of the following form:

$$K = 2\int_0^\infty \mathfrak{K}_\nu d\nu \tag{29}$$

where ν denotes the frequency of a color of wave length λ and whose velocity of propagation is q:

$$\nu = \frac{q}{\lambda},$$

and $\mathfrak{K}_\nu$ denotes the corresponding intensity of spectral radiation of the plane polarized light.

From the value of K is to be found the space density of radiation ϵ, i. e., the energy of radiation contained in unit volume. The point 0 in question forms the centre of a sphere whose radius r we take so small that in the distance r no appreciable absorption of radiation takes place. Then each element $d\sigma$ of the surface of the sphere furnishes, by virtue of the radiation traversing the same, the following contribution to the radiation density at 0:

$$\frac{d\sigma \cdot dt \cdot K \cdot d\Omega}{r^2 d\Omega \cdot qdt} = \frac{d\sigma \cdot K}{r^2 q}.$$

For the radiation cone of solid angle $d\Omega$ proceeding from a point of $d\sigma$ in the direction toward 0 has at the distance r from $d\sigma$ the cross-section $r^2 d\Omega$ and the energy passing in the time dt through this cross-section distributes itself along the distance qdt. By integration over all of the surface elements $d\sigma$ we obtain the total space density of radiation at 0:

$$\epsilon = \int \frac{d\sigma K}{r^2 q} = \frac{1}{q} \int K d\Omega,$$

wherein $d\Omega$ denotes the solid angle of an elementary cone whose vertex is 0. For uniform radiation we obtain:

$$\epsilon = \frac{4\pi K}{q} = \frac{8\pi}{q} \cdot \int_0^\infty \mathfrak{K}_\nu d\nu. \qquad (30)$$

The production of radiant heat is a consequence of the act of emission, and its destruction is the result of absorption. Both processes, emission and absorption, have their origin only in material particles, atoms or electrons, not at the geometrical bounding surface; although one frequently says, for the sake of brevity, that a surface element emits or absorbs. In reality a surface element of a body is a place of entrance for the radiation falling upon the body from without and which is to be absorbed; or a place of exit for the radiation emitted from within the body and passing through the surface in the outward direction. The capacity for emission and the capacity for absorption of an element of a body depend only upon its own condition and not upon that of the surrounding elements. If, therefore, as we shall assume in what follows, the state of the body varies only with the temperature, then the capacity for emission and the capacity for absorption of the body will also vary only with the temperature. The dependence upon the temperature can of course be different for each wave length.

We shall now introduce that result following from the second law of thermodynamics which will serve us as a basis in all subsequent considerations: "a system of bodies at rest of arbitrary nature, form and position, which is surrounded by a fixed shell impervious to heat, passes in the course of time from an arbitrarily chosen initial state to a permanent state in which the temperature of all bodies of the system is the same." This is the thermodynamic state of equilibrium in which the entropy of the system, among all those values which it may assume compatible with the total energy specified by the initial conditions, has a maximum value. Let us now apply this law to a single homogeneous isotropic medium which is of great extent in all directions of space and which, as in all cases subsequently considered, is surrounded by a fixed shell, perfectly reflecting as regards heat rays. The medium possesses for each frequency ν of the heat rays a finite capacity for emission and a finite capacity for absorption. Let us consider, now, such regions of the medium as are very far removed from the surface. Here the influence of the surface will be in any case vanishingly small, because no rays from the surface reach these regions, and on account of the homogeneity and isotropy of the medium we must conclude that the heat radiation is in thermodynamic equilibrium everywhere and has the same properties in all directions, so that $\mathfrak{K}_\nu$, the specific intensity of radiation of a plane polarized ray, is independent of the frequency ν, of the azimuth of polarization, of the direction of the ray, and of location. Thus, there will correspond to each diverging bundle of rays in an elementary cone $d\Omega$, proceeding from a surface element $d\sigma$, an exactly equal bundle oppositely directed, within the same elemental cone converging toward the surface element. This law retains its validity, as a simple consideration shows, right up to the surface of the medium, For in thermodynamic equilibrium each ray must possess exactly the same intensity as that of the directly opposite ray, otherwise, more energy would flow in one direction than in the opposite direction. Let us fix our attention upon a ray proceeding inwards from the surface, this must have the same intensity as that of the directly opposite ray coming from within, and from this it follows immediately that the state of radiation of the medium at all points on the surface is the same as that within. The nature of the bounding surface and the spacial extent of the medium are immaterial, and in a stationary state of radiation $\mathfrak{K}_\nu$ is completely determined by the nature of the medium for each temperature.

This law suffers a modification, however, in the special case that the medium is absolutely diathermanous for a definite frequency ν. It is then clear that the capacity for absorption and also that for emission must be zero, because otherwise no stationary state of radiation could exist, i. e., a medium emits no color which it does not absorb. But equilibrium can then obviously exist for every intensity of radiation of the frequency considered, i. e., $\mathfrak{K}_\nu$ is now undetermined and cannot be found without knowledge of the initial conditions. An important example of this is furnished by an absolute vacuum, which is diathermanous for all frequencies. In a complete vacuum thermodynamic equilibrium can therefore exist for each arbitrary intensity of radiation and for each frequency, i. e., for each arbitrary distribution of the spectral energy. From a general thermodynamic point of view this indeterminateness of the properties of thermodynamic states of equilibrium is explained through the presence of numerous different relative maxima of the entropy, as in the case of a vapor which is in a state of supersaturation. But among all the different maxima there is a special maximum, the absolute, which indicates stable equilibrium. In fact, we shall see that in a diathermanous medium for each temperature there exists a quite definite intensity of radiation, which is designated as the stable intensity of radiation of the frequency ν considered. But for the present we shall assume for all frequencies a finite capacity for absorption and for emission.

We consider now two homogeneous isotropic media in thermodynamic equilibrium separated from each other by a plane surface. Since the equilibrium will not be disturbed if one imagines for the moment the surface of separation between the two substances to be replaced by a surface quite non-transparent to heat radiation, all of the foregoing laws hold for each of the

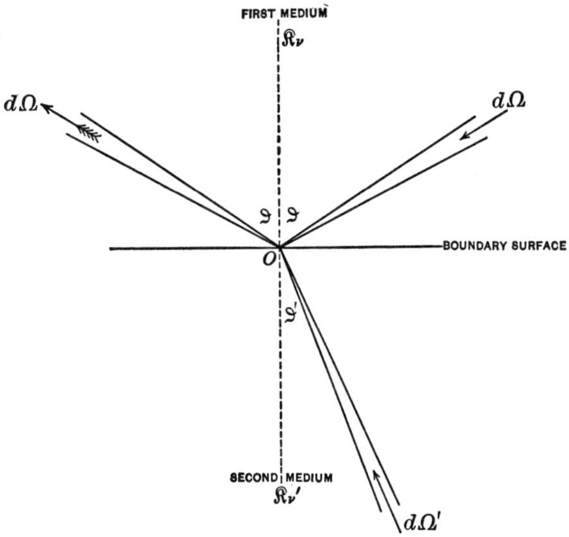

Fig. 1.

two substances individually. Let the specific intensity of radiation of frequency ν, polarized in any arbitrary plane within the first substance (the upper in Fig. 1)[1], be $\mathfrak{K}_\nu$ and that within the second substance $\mathfrak{K}_\nu'$ (we shall in general designate with a dash

[1] From my lectures upon the theory of heat radiation (Leipzig, J. A. Barth), wherein are to be found the details of the above somewhat abbreviated calculations.

those quantities which refer to the second substance). Both quantities $\mathfrak{K}_\nu$ and $\mathfrak{K}_\nu'$, besides depending upon the temperature and the frequency, depend only upon the nature of the two substances, and, in fact, these values of the intensity of radiation hold quite up to the boundary surface between the substances, and are therefore independent of the properties of this surface.

Each ray from the first medium is split into two rays at the boundary surface: the reflected and the transmitted. The directions of these two rays vary according to the angle of incidence and the color of the incident ray, and, in addition, the intensity varies according to its polarization. If we denote by ρ (the reflection coefficient) the amount of the reflected energy of radiation and consequently by $1 - \rho$ the amount of transmitted energy with respect to the incident energy, then ρ depends upon the angle of incidence, upon the frequency and upon the polarization of the incident ray. Similar remarks hold for ρ', the reflection coefficient for a ray from the second medium, upon meeting the boundary surface.

Now the energy of a monochromatic plane polarized ray of frequency ν proceeding from an element $d\sigma$ of the boundary surface within the elementary cone $d\Omega$ in a direction toward the first medium (see the feathered arrow at the left in Fig. 1) is for the time dt, in accordance with (28) and (29):

$$dt \cdot d\sigma \cdot \cos\vartheta \cdot d\Omega \cdot \mathfrak{K}_\nu d\nu, \tag{31}$$

where

$$d\Omega = \sin\vartheta d\vartheta d\varphi. \tag{32}$$

This energy is furnished by the two rays which, approaching the surface from the first and the second medium respectively, are reflected and transmitted respectively at the surface element $d\sigma$ in the same direction. (See the unfeathered arrows. The surface element $d\sigma$ is indicated only by the point 0.) The first ray proceeds in accordance with the law of reflection within the symmetrically drawn elementary cone $d\Omega$: the second approaches the surface within the elementary cone

$$d\Omega' = \sin \vartheta' d\vartheta' d\varphi', \tag{33}$$

where, in accordance with the law of refraction,

$$\varphi' = \varphi \quad \text{and} \quad \frac{\sin \vartheta}{\sin \vartheta'} = \frac{q}{q'}. \tag{34}$$

We now assume that the ray is either polarized in the plane of incidence or perpendicular to this plane, and likewise for the two radiations out of whose energies it is composed. The radiation coming from the first medium and reflected from $d\sigma$ contributes the energy:

$$\rho \cdot dt \cdot d\sigma \cos \vartheta \cdot d\Omega \cdot \mathfrak{K}_\nu d\nu, \tag{35}$$

and the radiation coming from the second medium and transmitted through $d\sigma$ contributes the energy:

$$(1 - \rho') \cdot dt \cdot d\sigma \cos \vartheta' \cdot d\Omega' \cdot \mathfrak{K}_\nu' d\nu. \tag{36}$$

The quantities dt, $d\sigma$, ν, and $d\nu$ are here written without the accent, since they have the same values in both media.

Adding the expressions (35) and (36) and placing the sum equal to the expression (31), we obtain:

$$\rho \cos \vartheta d\Omega \mathfrak{K}_\nu + (1 - \rho') \cos \vartheta' d\Omega' \mathfrak{K}_\nu' = \cos \vartheta d\Omega \mathfrak{K}_\nu.$$

Now, in accordance with (34):

$$\frac{\cos \vartheta d\vartheta}{q} = \frac{\cos \vartheta' d\vartheta'}{q'},$$

and further, taking note of (32) and (33):

$$d\Omega' \cos \vartheta' = d\Omega \cos \vartheta \cdot \frac{q'^2}{q^2},$$

and it follows that:

$$\rho \mathfrak{K}_\nu + (1 - \rho') \frac{q'^2}{q^2} \mathfrak{K}_\nu' = \mathfrak{K}_\nu$$

or:

$$\frac{\mathfrak{K}_\nu}{\mathfrak{K}_\nu'} \cdot \frac{q^2}{q'^2} = \frac{1 - \rho'}{1 - \rho}.$$

In the last equation the quantity on the left is independent of the angle of incidence ϑ and of the kind of polarization, consequently the quantity upon the right side must also be independent of these quantities. If one knows the value of these quantities for a single angle of incidence and for a given kind of polarization, then this value is valid for all angles of incidence and for all polarizations. Now, in the particular case that the rays are polarized at right angles to the plane of incidence and meet the bounding surface at the angle of polarization,

$$\rho = 0 \quad \text{and} \quad \rho' = 0.$$

Then the expression on the right will be equal to 1, and therefore it is in general equal to 1, and we have always:

$$\rho = \rho', \quad q^2\mathfrak{K}_\nu = q'^2\mathfrak{K}_\nu'. \tag{37}$$

The first of these two relations, which asserts that the coefficient of reflection is the same for both sides of the boundary surface, constitutes the special expression of a general reciprocal law, first announced by Helmholz, whereby the loss of intensity which a ray of given color and polarization suffers on its path through any medium in consequence of reflection, refraction, absorption, and dispersion is exactly equal to the loss of intensity which a ray of corresponding intensity, color and polarization suffers in passing over the directly opposite path. It follows immediately from this that the radiation meeting a boundary surface between two media is transmitted or reflected equally well from both sides, for every color, direction and polarization.

The second relation, (37), brings into connection the radiation intensities originating in both substances. It asserts that in thermodynamic equilibrium the specific intensities of radiation of a definite frequency in both media vary inversely as the square of the velocities of propagation, or directly as the squares of the refractive indices. We may therefore write

$$q^2\mathfrak{K}_\nu = F(\nu, T),$$

wherein F denotes a universal function depending only upon ν and T, the discovery of which is one of the chief problems of the theory.

Let us fix our attention again on the case of a diathermanous medium. We saw above that in a medium surrounded by a non-transparent shell which for a given color is diathermanous equilibrium can exist for any given intensity of radiation of this color. But it follows from the second law that, among all the intensities of radiation, a definite one, namely, that corresponding to the absolute maximum of the total entropy of the system, must exist, which characterizes the absolutely stable equilibrium of radiation. We now see that this indeterminateness is eliminated by the last equation, which asserts that in thermodynamic equilibrium the product $q^2\mathfrak{K}_\nu$ is a universal function. For it results immediately therefrom that there is a definite value of $\mathfrak{K}_\nu$ for every diathermanous medium which is thus differentiated from all other values. The physical meaning of this value is derived directly from a consideration of the way in which this equation was derived: it is that intensity of radiation which exists in the diathermanous medium when it is in thermodynamic equilibrium while in contact with a given absorbing and emitting medium. The volume and the form of the second medium is immaterial; in particular, the volume may be taken arbitrarily small.

For a vacuum, the most diathermanous of all media, in which the velocity of propagation $q = c$ is the same for all rays, we can therefore express the following law: The quantity

$$\mathfrak{K}_\nu = \frac{1}{c^2} F(\nu, T) \tag{38}$$

denotes that intensity of radiation which exists in any complete vacuum when it is in a stationary state as regards exchange of radiation with any absorbing and emitting substance, whose amount may be arbitrarily small. This quantity $\mathfrak{K}_\nu$ regarded as a function of ν gives the so-called normal energy spectrum.

Let us consider, therefore, a vacuum surrounded by given emitting and absorbing bodies of uniform temperature. Then, in the course of time, there is established therein a normal energy radiation $\mathfrak{K}_\nu$ corresponding to this temperature. If now ρ_ν be the reflection coefficient of a wall for the frequency ν, then of the radiation $\mathfrak{K}_\nu$ falling upon the wall, the part $\rho_\nu \mathfrak{K}_\nu$ will be reflected. On the other hand, if we designate by E_ν the emission coefficient of the wall for the same frequency ν, the total radiation proceeding from the wall will be:

$$\rho_\nu \mathfrak{K}_\nu + E_\nu = \mathfrak{K}_\nu,$$

since each bundle of rays possesses in a stationary state the intensity $\mathfrak{K}_\nu$. From this it follows that:

$$\mathfrak{K}_\nu = \frac{E_\nu}{1 - \rho_\nu}, \tag{39}$$

i. e., the ratio of the emission coefficient E_ν to the capacity for absorption $(1 - \rho_\nu)$ of a given substance is the same for all substances and equal to the normal intensity of radiation for each frequency (Kirchoff). For the special case that ρ_ν is equal to 0, i. e., that the wall shall be perfectly black, we have:

$$\mathfrak{K}_\nu = E_\nu,$$

that is, the normal intensity of radiation is exactly equal to the emission coefficient of a black body. Therefore the normal radiation is also called "black radiation." Again, for any given body, in accordance with (39), we have:

$$E_\nu < \mathfrak{K}_\nu,$$

i. e., the emission coefficient of a body in general is smaller than that of a black body. Black radiation, thanks to W. Wien and O. Lummer, has been made possible of measurement, through a small hole bored in the wall bounding the space considered.

We proceed now to the treatment of the problem of determining the specific intensity $\mathfrak{K}_\nu$ of black radiation in a vacuum, as regards its dependence upon the frequency ν and the temperature T. In the treatment of this problem it will be necessary to go further than we have previously done into those processes which condition the production and destruction of heat rays; that is, into the question regarding the act of emission and that of absorption. On account of the complicated nature of these processes and the difficulty of bringing some of the details into connection with experience, it is certainly quite out of the question to obtain in this manner any reliable results if the following law cannot be utilized as a dependable guide in this domain: a vacuum surrounded by reflecting walls in which arbitrary emitting and absorbing bodies are distributed in any given arrangement assumes in the course of time the stationary state of black radiation, which is completely determined by a single parameter, the temperature, and which, in particular, does not depend upon the number, the properties and the arrangement of the bodies. In the investigation of the properties of the state of black radiation the nature of the bodies which are supposed to be in the vacuum is therefore quite immaterial, and it is certainly immaterial whether such bodies actually exist anywhere in nature, so long as their existence and their properties are compatible throughout with the laws of electrodynamics and of thermodynamics. As soon as it is possible to associate with any given special kind and arrangement of emitting and absorbing bodies a state of radiation in the surrounding vacuum which is characterized by absolute stability, then this state can be no other than that of black radiation. Making use of the freedom furnished by this law, we choose among all the emitting and absorbing systems conceivable, the most simple, namely, a single oscillator at rest, consisting of two poles charged with equal quantities of electricity of opposite sign which are movable relative to each other in a fixed straight line, the axis of the oscillator. The state of the oscillator is completely determined by its moment $f(t)$; i. e., by the product of the electric charge of the pole on the positive side of the axis into the distance between the poles, and by its differential quotient with regard to the time:

$$\frac{df(t)}{dt} = \dot{f}(t).$$

The energy of the oscillator is of the following simple form:

$$U = \tfrac{1}{2}Kf^2 + \tfrac{1}{2}L\dot{f}^2, \tag{40}$$

wherein K and L denote positive constants which depend upon the nature of the oscillator in some manner into which we need not go further at this time.

If, in the vibrations of the oscillator, the energy U remain absolutely constant, we should have: $dU = 0$ or:

$$Kf(t) + L\ddot{f}(t) = 0,$$

and from this there results, as a general solution of the differential equation, a pure periodic vibration:

$$f = C \cos (2\pi\nu_0 t - \vartheta),$$

wherein C and ϑ denote the integration constants and ν_0 the number of vibrations per unit of time:

$$\nu_0 = \frac{1}{2\pi}\sqrt{\frac{K}{L}}. \tag{41}$$

Such an oscillator vibrating periodically with constant energy would neither be influenced by the electromagnetic field surrounding it, nor would it exert any external actions due to radiation. It could therefore have no sort of influence on the heat radiation in the surrounding vacuum.

In accordance with the theory of Maxwell, the energy of vibration U of the oscillator by no means remains constant in general, but an oscillator by virtue of its vibrations sends out spherical waves in all directions into the surrounding field and, in accordance with the principle of conservation of energy, if no actions from without are exerted upon the oscillator, there must necessarily be a loss in the energy of vibration and, therefore, a damping of the amplitude of vibration is involved. In order to find the amount of this damping we calculate the quantity of energy which flows out through a spherical surface with the oscillator at the center, in accordance with the law of Poynting. However, we may not place the energy flowing outwards in accordance with this law through the spherical surface in an infinitely small interval of time *dt* equal to the energy radiated in the same time interval from the oscillator. For, in general, the electromagnetic energy does not always flow in the outward direction, but flows alternately outwards and inwards, and we should obtain in this manner for the quantity of the radiation outwards, values which are alternately positive and negative, and which also depend essentially upon the radius of the supposed sphere in such manner that they increase toward infinity with decreasing radius—which is opposed to the fundamental conception of radiated energy. This energy will, moreover, be only found independent of the radius of the sphere when we calculate the total amount of energy flowing outwards through the surface of the sphere, not for the time element *dt*, but for a sufficiently large time. If the vibrations are purely periodic, we may choose for the time a period; if this is not the case, which for the sake of generality we must here assume, it is not possible to specify a priori any more general criterion for the least possible necessary magnitude of the time than that which makes the energy radiated essentially independent of the radius of the supposed sphere.

In this way we succeed in finding for the energy emitted from the oscillator in the time from t to $t + \mathfrak{T}$ the following expression:

$$\frac{2}{3c^3}\int_t^{t+\mathfrak{T}} \ddot{f}^2(t)dt.$$

If now, the oscillator be in an electromagnetic field which has the electric component $\mathfrak{E}_z$ at the oscillator in the direction of its axis, then the energy absorbed by the oscillator in the same time is:

$$\int_t^{t+\mathfrak{T}} \mathfrak{E}_z \dot{f} \cdot dt.$$

Hence, the principle of conservation of energy is expressed in the following form:

$$\int_t^{t+\mathfrak{T}} \left(\frac{dU}{dt} + \frac{2}{3c^3} \ddot{f}^2 - \mathfrak{E}_z \dot{f} \right) dt = 0.$$

This equation, together with the assumption that the constant

$$\frac{4\pi^2 \nu_0}{3c^3 L} = \sigma \tag{42}$$

is a small number, leads to the following linear differential equation for the vibrations of the oscillator:

$$Kf + L\ddot{f} - \frac{2}{3c^3} \dddot{f} = \mathfrak{E}_z. \tag{43}$$

In accordance with what precedes, in so far as the oscillator is excited into vibrations by an external field $\mathfrak{E}_z$, one may designate it as a resonator which possesses the natural period ν_0 and the small logarithmic decrement σ. The same equation may be obtained from the electron theory, but I have considered it an advantage to derive it in a manner independent of any hypothesis concerning the nature of the resonator.

Now, let the resonator be in a vacuum filled with stationary black radiation of specific intensity $\mathfrak{K}_\nu$. How, then, does the mean energy U of the resonator in a state of stationary vibration depend upon the specific intensity of radiation $\mathfrak{K}_{\nu_0}$ with the natural period ν_0 of the corresponding color? It is this question which we have still to consider today. Its answer will be found by expressing on the one hand the energy of the resonator U and on the other hand the intensity of radiation $\mathfrak{K}_{\nu_0}$ by means of the component $\mathfrak{E}_z$ of the electric field exciting the resonator. Now however complicated this quantity may be, it is capable of development in any case for a very large time interval, from $t = 0$ to $t = \mathfrak{T}$, in the Fourier's series:

$$\mathfrak{E}_z = \sum_{n=1}^{n=\infty} C_n \cos\left(\frac{2\pi nt}{\mathfrak{T}} - \vartheta_n\right), \tag{44}$$

and for this same time interval $\mathfrak{T}$ the moment of the resonator in the form of a Fourier's series may be calculated as a function of t from the linear differential equation (43). The initial condition of the resonator may be neglected if we only consider such times t as are sufficiently far removed from the origin of time $t = 0$.

If it be now recalled that in a stationary state of vibration the mean energy U of the resonator is given, in accordance with (40), (41) and (42), by:

$$U = K\overline{f^2} = \frac{16\pi^4\nu_0^3}{3\sigma c^3}\overline{f^2},$$

it appears atter substitution of the value of f obtained from the differential equation (43) that:

$$U = \frac{3c^3}{64\pi^2\nu_0^2}\mathfrak{T}\overline{C_{n0}^2}, \tag{45}$$

wherein $\overline{C_{n0}^2}$ denotes the mean value of C_n for all the series of numbers n which lie in the neighborhood of the value $\nu_0\mathfrak{T}$, i. e., for which $\nu_0\mathfrak{T}$ is approximately $= 1$.

Now let us consider on the other hand the intensity of black radiation, and for this purpose proceed from the space density of the total radiation. In accordance with (30), this is:

$$\epsilon = \frac{8\pi}{c}\int_0^\infty \mathfrak{K}_\nu d\nu = \frac{1}{8\pi}(\overline{\mathfrak{E}_x^2} + \overline{\mathfrak{E}_y^2} + \overline{\mathfrak{E}_z^2} + \overline{\mathfrak{H}_x^2} + \overline{\mathfrak{H}_y^2} + \overline{\mathfrak{H}_z^2}), \tag{46}$$

and therefore, since the radiation is isotropic, in accordance with (44):

$$\frac{8\pi}{c}\int_0^\infty \mathfrak{K}_\nu d\nu = \frac{3}{4\pi}\overline{\mathfrak{E}_z^2} = \frac{3}{8\pi}\sum_{n=1}^{n=\infty} C_n^2.$$

If we write $\Delta n/\mathfrak{T}$ on the left instead of $d\nu$, where Δn is a large number, we get:

$$\frac{8\pi}{c}\sum_{n=1}^{n=\infty}\mathfrak{K}_\nu\frac{\Delta n}{\mathfrak{T}}=\frac{3}{8\pi}\sum_{n=1}^{n=\infty}C_n^2,$$

and obtain then by "spectral" division of this equation:

$$\frac{8\pi}{c}\mathfrak{K}_{\nu 0}\frac{\Delta n}{\mathfrak{T}}=\frac{3}{8\pi}\sum_{n_0-(\Delta n/2)}^{n_0+(\Delta n/2)}C_n^2,$$

and, if we introduce again the mean value

$$\frac{1}{\Delta n}\cdot\sum_{n_0-(\Delta n/2)}^{n_0+(\Delta n/2)}C_n^2=\overline{C}_{n0}^2,$$

we then get:

$$\mathfrak{K}_{\nu 0}=\frac{3c\mathfrak{T}}{64\pi^2}\cdot\overline{C}_{n0}.$$

By comparison with (45) the relation sought is now found:

$$\mathfrak{K}_{\nu 0}=\frac{\nu_0^2}{c^2}U, \tag{47}$$

which is striking on account of its simplicity and, in particular, because it is quite independent of the damping constant σ of the resonator.

This relation, found in a purely electrodynamic manner, between the spectral intensity of black radiation and the energy of a vibrating resonator will furnish us in the next lecture, with the aid of thermodynamic considerations, the necessary means of attack in deriving the temperature of black radiation together with the distribution of energy in the normal spectrum.

SIXTH LECTURE.

HEAT RADIATION. STATISTICAL THEORY.

Following the preparatory considerations of the last lecture we shall treat today the problem which we have come to recognize as one of the most important in the theory of heat radiation: the establishment of that universal function which governs the energy distribution in the normal spectrum. The means for the solution of this problem will be furnished us through the calculation of the entropy S of a resonator placed in a vacuum filled with black radiation and thereby excited into stationary vibrations. Its energy U is then connected with the corresponding specific intensity $\mathfrak{K}_\nu$ and its natural frequency ν in the radiation of the surrounding field through equation (47):

$$\mathfrak{K}_\nu = \frac{\nu^2}{c^2} U. \tag{48}$$

When S is found as a function of U, the temperature T of the resonator and that of the surrounding radiation will be given by:

$$\frac{dS}{dU} = \frac{1}{T}, \tag{49}$$

and by elimination of U from the last two equations, we then find the relationship among $\mathfrak{K}_\nu$, T and ν.

In order to find the entropy S of the resonator we will utilize the general connection between entropy and probability, which we have extensively discussed in the previous lectures, and inquire then as to the existing probability that the vibrating resonator possesses the energy U. In accordance with what we have seen in connection with the elucidation of the second law through atomistic ideas, the second law is only applicable to a physical system when we consider the quantities which determine the state of the system as mean values of numerous disordered individual values, and the probability of a state is then equal to the number of the numerous, a priori equally probable, complexions which make possible the realization of the state. Accordingly, we have to consider the energy U of a resonator placed in a stationary field of black radiation as a constant mean value of many disordered independent individual values, and this procedure agrees with the fact that every measurement of the intensity of heat radiation is extended over an enormous number of vibration periods. The entropy of a resonator is then to be calculated from the existing probability that the energy of the radiator possesses a definite mean value U within a certain time interval.

In order to find this probability, we inquire next as to the existing probability that the resonator at any fixed time possesses a given energy, or in other words, that that point (the state point) which through its coordinates indicates the state of the resonator falls in a given "state domain." At the conclusion of the third lecture (p. 57) we saw in general that this probability is simply measured through the magnitude of the corresponding state domain:

$$\int d\varphi \cdot d\psi,$$

in case one employs as coordinates of state the general coordinate φ and the corresponding momentum ψ. Now in general, the energy of the resonator, in accordance with (40), is:

$$U = \tfrac{1}{2}Kf^2 + \tfrac{1}{2}L\dot{f}^2.$$

If we choose f as the general coordinate φ and put, therefore, $\varphi = f$, then the corresponding impulse ψ is equal

$$\frac{\partial U}{\partial \dot{f}} = L\dot{f},$$

and the energy U expressed as a function of φ and ψ is:

$$U = \tfrac{1}{2}K\varphi^2 + \frac{1}{2}\frac{\psi^2}{L}.$$

If now we desire to find the existing probability that the energy of a resonator shall lie between U and $U + \Delta U$, we have to calculate the magnitude of that state domain in the (φ, ψ)-plane which is bounded by the curves $U =$ const. and $U + \Delta U =$ const. These two curves are similar and similarly placed ellipses and the portion of surface bounded by them is equal to the difference of the areas of the two ellipses. The areas are respectively U/ν and $(U + \Delta U)/\nu$; consequently, the magnitude sought for the state domain is: $\Delta U/\nu$. Let us now consider the whole state plane so divided into elementary portions by a large number of ellipses, such that the annular areas between consecutive ellipses are equal to each other; i. e., so that:

$$\frac{\Delta U}{\nu} = \text{const} = h.$$

We thus obtain those portions ΔU of the energy which correspond to equal probabilities and which are therefore to be designated as the energy elements:

$$\epsilon = \Delta U = h\nu. \tag{50}$$

If the determination of the elementary domains is effected in a manner quite similar to that employed in the kinetic gas theory, there exist, with respect to the relationships there found, very notable differences. In the first place, the state of the physical system considered here, the resonator, does not depend as there upon the coordinates and the velocities, but upon the energy only, and this circumstance necessitates that the entropy of a state depend, not upon the distribution of the state quantities φ and ψ, but only upon the energy U. A further difference consists in this, that we have to do in the case of molecules with spacial mean values, but in the case of radiation with mean values as regards time. But this distinction may be disregarded when we reflect that the mean time value of the energy U of a given resonator is obviously identical with the mean space value at a given instant of time of a great number N of similar resonators distributed in the same stationary field of radiation. Of course these resonators must be placed sufficiently far apart in order not directly to influence one another. Then the total energy of all the resonators:

$$U_N = NU \tag{51}$$

is quite irregularly distributed among all the individual resonators, and we have referred back the disorder as regards time to a disorder as regards space.

We are now concerned with the probability W of the state determined by the energy U_N of the N resonators placed in the same stationary field of radiation; i. e., with the number of individual arrangements or complexions which correspond to the distribution of energy U_N among the N resonators. With this in view, we subdivide the given total energy U_N into its elements ϵ so that:

$$U_N = P\epsilon. \tag{52}$$

These P energy elements are to be distributed in every possible manner among the N resonators. Let us consider, then, the N resonators to be numbered and the figures written beside one another in a series, and in such manner that the number of times each figure appears is equal to the number of energy elements which fall upon the corresponding resonator. Then we obtain through such a number series a representation of a fixed complexion, in which with each individual resonator there is associated a definite energy. For example, if there are $N = 4$ resonators and $P = 6$ energy elements present, then one of the possible complexions is represented by the number series

1 1 3 3 3 4

which asserts that the first resonator contains two, the second 0, the third 3, and the fourth 1 energy element. The totality of numbers in the series is 6, equal to the number of the energy elements present. The arrangement of figures in the series is immaterial for any complexion, since the mere interchange of figures does not change the energy of a given resonator. The number of all the possible different complexions is therefore equal to the number of possible "combinations with repetition" of 4 elements with 6 classes:

$$W = \frac{(4+6-1)!}{(4-1)!6!} = \frac{9!}{3!6!} = 84,$$

or, in our general case the probability sought is:

$$W = \frac{(N+P-1)!}{(N-1)!P!}.$$

We obtain, therefore, for the entropy S_N of the resonator system, in accordance with equation (12), since N and P are large numbers,

$$S_N = k \log \frac{(N+P)!}{N!P!}$$

and with the aid of Sterling's formula (16):

$$S_N = k\{(N+P) \log (N+P) - N \log N - P \log P\}.$$

If, in accordance with (52), we now write U_N/ϵ for P, NU for U_N in accordance with (51), and $h\nu$ for ϵ, in accordance with (50), we obtain, after an easy transformation, for the mean entropy of a single resonator:

$$\frac{S_N}{N} = S = k\left\{\left(1+\frac{U}{h\nu}\right)\log\left(1+\frac{U}{h\nu}\right) - \frac{U}{h\nu}\log\frac{U}{h\nu}\right\}$$

as the solution of the problem in hand.

We will now introduce the temperature T of the resonator, and will express through T the energy U of the resonator and also the intensity $\mathfrak{K}_\nu$ of the heat radiation related to it through a stationary state of energy exchange. For this purpose we utilize equation (49) and obtain then for the energy of the resonator:

$$U = \frac{h\nu}{e^{h\nu/kT} - 1}.$$

It is to be observed that we have not here to do with a uniform distribution of energy (cf. p. 68) among the various resonators.

For the specific intensity of the monochromatic plane polarized ray of frequency ν, we have, in accordance with (48):

$$\mathfrak{K}_\nu = \frac{h\nu^3}{c^2} \cdot \frac{1}{e^{h\nu/kT} - 1}. \tag{53}$$

This expression furnishes for each temperature T the energy distribution in the normal spectrum of a black body. A comparison with equation (38) of the last lecture furnishes us then with the universal function:

$$F(\nu, T) = \frac{h\nu^3}{e^{h\nu/kT} - 1}.$$

If we refer the specific intensity of a monochromatic ray, not to the frequency ν, but, as is commonly done in experimental physics, to the wave length λ, then, since between the absolute values of $d\nu$ and $d\lambda$ the relation exists:

$$|d\nu| = \frac{c \cdot |d\lambda|}{\lambda^2},$$

we obtain from

$$E_\lambda |d\lambda| = \mathfrak{K}_\nu |d\nu|,$$

the relation:

$$E_\lambda = \frac{c^2 h}{\lambda^5} \cdot \frac{1}{e^{ch/k\lambda T} - 1} \tag{54}$$

as the intensity of a monochromatic plane polarized ray of wave length λ which is emitted normally to the surface of a black body in a vacuum at temperature T. For small values of λT

(54) reduces to:

$$E_\lambda = \frac{c^2 h}{\lambda^5} \cdot e^{-(ch/k\lambda T)}, \tag{55}$$

which expresses Wien's Displacement Law. For large values of λT on the other hand, there results from (54):

$$E_\lambda = \frac{ckT}{\lambda^4}, \tag{56}$$

a relation first established by Lord Rayleigh and which we may here designate as the Rayleigh Law of Radiation.

From equation (30), taking account of (53), we obtain for the space density of black radiation in a vaccuum:

$$\epsilon = \frac{48\pi h}{c^3}\left(\frac{kT}{h}\right)^4 \cdot \alpha = aT^4,$$

wherein

$$\alpha = 1 + \frac{1}{2^4} + \frac{1}{3^4} + \frac{1}{4^4} + \cdots = 1.0823.$$

The Stefan-Boltzmann law is hereby expressed. In accordance with the measurements of Kurlbaum, we have the constant

$$a = \frac{48\pi k^4}{c^3 h^3} \cdot \alpha = 7.061 \cdot 10^{-15} \frac{\text{erg}}{\text{cm}^3\,\text{deg}^4}.$$

For that wave length λ_m which corresponds in the spectrum of black radiation to the maximum intensity of radiation E_λ we have from equation (54):

$$\left(\frac{dE_\lambda}{d\lambda}\right)_{\lambda=\lambda_m} = 0.$$

Carrying out the differentiation, we get, after putting for brevity:

$$\frac{ch}{k\lambda_m T} = \beta, \quad e^{-\beta} + \frac{\beta}{5} - 1 = 0.$$

The root of this transcendental equation is:

$$\beta = 4.9651;$$

and $\lambda_m T = ch/k\beta = b$ is a constant (Wien's Displacement Law). In accordance with the measurements of O. Lummer and E. Pringsheim,

$$b = 0.294 \text{ cm} \cdot \text{deg}.$$

From this there follow the numerical values

$$k = 1.346 \cdot 10^{-16} \frac{\text{erg}}{\text{deg}}, \quad \text{and} \quad h = 6.548 \cdot 10^{-27} \text{ erg} \cdot \text{sec}.$$

The value found for k easily permits of the specification numerically, in the C.G.S. system, of the general connection between entropy and probability, as expressed through the universal equation (12). Thus, quite in general, the entropy of a physical system is:

$$S = 1.346 \cdot 10^{-16} \overset{e}{\log} W.$$

In the application to the kinetic gas theory we obtain from equation (24) for the ratio of the molecular mass to the mol mass:

$$\omega = \frac{k}{R} = 1.62 \cdot 10^{-24},$$

i. e., to one mol there corresponds $1/\omega = 6.175 \cdot 10^{23}$ molecules, where it is supposed that the mol of oxygen

$$O_2 = 32\text{g}.$$

Accordingly, the number of molecules contained in 1 cu. cm. of an ideal gas at 0° Cels. and at atmospheric pressure is:

$$N = 2.76 \cdot 10^{19}.$$

The mean kinetic energy of the progressive motion of a molecule at the absolute temperature $T = 1$ in the absolute C.G.S. system, in accordance with (27), is:

$$L = \tfrac{3}{2}k = 2.02 \cdot 10^{-16}.$$

In general, the mean kinetic energy of progressive motion of a molecule is expressed by the product of this number and the absolute temperature T.

The elementary quantum of electricity, or the free electric charge of a monovalent ion or electron, in electrostatic measure is:

$$e = \infty \cdot 9658 \cdot 3 \cdot 10^{10} = 4.69 \cdot 10^{-10}.$$

This result stands in noteworthy agreement with the results of the latest direct measurements of the electric elementary quantum made by E. Rutherford and H. Geiger, and E. Regener.—

Even if the radiation formula (54) here derived had shown itself as valid with respect to all previous tests, the theory would still require an extension as regards a certain point; for in it the physical meaning of the universal constant h remains quite unexplained. All previous attempts to derive a radiation formula upon the basis of the known laws of electron theory, among which the theory of J. H. Jeans is to be considered as the most general and exact, have led to the conclusion that h is infinitely small, so that, therefore, the radiation formula of Rayleigh possesses general validity, but, in my opinion, there can be no doubt that this formula loses its validity for short waves, and that the pains which Jeans has taken to place[1] the blame for the contradiction between theory and experiment upon the latter are unwarranted.

Consequently, there remains only the one conclusion, that previous electron theories suffer from an essential incompleteness which demands a modification, but how deeply this modification should go into the structure of the theory is a question upon which views are still widely divergent. J. J. Thompson inclines to the most radical view, as do J. Larmor, A. Einstein, and with him I. Stark, who even believe that the propagation of electromagnetic waves in a pure vacuum does not occur precisely in accordance with the Maxwellian field equations, but in definite energy quanta $h\nu$. I am of the opinion, on the other hand, that at present it is not necessary to proceed in so revolu-

[1] In that the walls used in the measurements of hollow space radiations must be diathermanous for the shortest waves.

tionary a manner, and that one may come successfully through by seeking the significance of the energy quantum $h\nu$ solely in the mutual actions with which the resonators influence one another.[1] A definite decision with regard to these important questions can only be brought about as a result of further experience.

[1] It is my intention to give a complete presentation of these relations in Volume 31 of the Annalen der Physik.

SEVENTH LECTURE.

General Dynamics. Principle of Least Action.

Since I began three weeks ago today to depict for you the present status of the system of theoretical physics and its probable future development, I have continually sought to bring out that in the theoretical physics of the future the most important and the final division of all physical processes would likely be into reversible and irreversible processes. In succeeding lectures, with the aid of the calculus of probability and with the introduction of the hypothesis of elementary disorder, we have seen that all irreversible processes may be considered as reversible elementary processes: in other words, that irreversibility does not depend upon an elementary property of a physical process, but rather depends upon the ensemble of numerous disordered elementary processes of the same kind, each one of which individually is completely reversible, and upon the introduction of the macroscopic method of treatment. From this standpoint one can say quite correctly that in the final analysis all processes in nature are reversible. That there is herein contained no contradiction to the principle regarding the irreversibility of processes expressed in terms of the mean values of elementary processes of macroscopic changes of state, I have demonstrated fully in the third lecture. Perhaps it will be appropriate at this place to interject a more general statement. We are accustomed in physics to seek the explanation of a natural process by the method of division of the process into elements. We regard each complicated process as composed of simple elementary processes, and seek to analyse it through thinking of the whole as the sum of the parts. This method, however, presupposes that through this division the character of the whole is not changed; in somewhat similar manner each measurement of a physical process presupposes that the progress of the phenomena is not influenced by the introduction of the measuring instrument. We have here a case in which that supposition is not warranted, and where a direct conclusion with regard to the parts applied to the whole leads to quite false results. If we divide an irreversible process into its elementary constituents, the disorder and along with it the irreversibility vanishes; an irreversible process must remain beyond the understanding of anyone who relies upon the fundamental law: that all properties of the whole must also be recognizable in the parts. It appears to me as though a similar difficulty presents itself in most of the problems of intellectual life.

Now after all the irreversibility in nature thus appears in a certain sense eliminated, it is an illuminating fact that general elementary dynamics has only to do with reversible processes. Therefore we shall occupy ourselves in what follows with reversible processes exclusively. That which makes this procedure so valuable for the theory is the circumstance that all known reversible processes, be they mechanical, electrodynamical or thermal, may be brought together under a single principle which answers unambiguously all questions regarding their behavior. This principle is not that of conservation of energy; this holds, it is true, for all these processes, but does not determine unambiguously their behavior; it is the more comprehensive principle of least action.

The principle of least action has grown upon the ground of mechanics where it enjoys equal rank and regard with numerous other principles; the principle of d'Alembert, the principle of virtual displacement, Gauss's principle of least constraint, the Lagrangian Equations of the first and second kind. All these principles are equivalent to one another and therefore at bottom are only different formularizations of the same laws; sometimes one and sometimes another is the most convenient to use. But the principle of least action has the decided advantage over all the other principles mentioned in that it connects together in a single equation the relations between quantities which possess, not only for mechanics, but also for electrodynamics and for thermodynamics, direct significance, namely, the quantities: space, time and potential. This is the reason why one may directly apply the principle of least action to processes other than mechanical, and the result has shown that such applications, as well in electrodynamics as in thermodynamics, lead to the appropriate laws holding in these subjects. Since a representation of a unified system of theoretical physics such as we have here in mind must lay the chief emphasis upon as general an interpretation as possible of physical laws, it is self evident that in our treatment the principle of least action will be called upon to play the principal rôle. I desire now to show how it is applied in simple individual cases.

The general formularization of the principle of least action in the interpretation given to it by Helmholz is as follows: among all processes which may carry a certain arbitrarily given physical system subject to given external actions from a given initial position into a given final position in a given time, the process which actually takes place in nature is that which is distinguished by the condition that the integral

$$\int_{t_0}^{t_1} (\delta H + A)dt = 0, \tag{57}$$

wherein an arbitrary displacement of the independent coordinates (and velocities) is denoted by the sign δ, and A denotes the infinitely small increase in energy (external work) which the system experiences in the displacement δ. The function H is the kinetic potential. When we speak here of the positions, the coordinates, and the velocities of the configuration, we understand thereby, not only those special ones corresponding to mechanical ideas, but also all the so-called generalized coordinates with the quantities derived therefrom; and these may represent equally well quantities of electricity, volumes, and the like.

In the applications which we shall now make of the principle of least action, we must first decide as to whether the generalized coordinates which determine the state of the system considered are present in finite number or form a continuous infinite manifold. We shall distinguish the examples here considered in accordance with this viewpoint.

1. *The Position (Configuration) is Determined by a Finite Number of Coordinates.*

In ordinary mechanics this is actually the case in every system of a finite number of material points or rigid bodies among whose coordinates there exist arbitrary fixed equations of condition. If we call the independent coordinates $\varphi_1, \varphi_2, \cdots$, then the external work is:

$$A = \Phi_1 \delta\varphi_1 + \Phi_2 \delta\varphi_2 + \cdots = \delta E, \tag{58}$$

wherein $\Phi_1, \Phi_2, \cdots$ are the "external force components" which correspond to the individual coordinates, and E denotes the energy of the system. Then the principle of least action is expressed by:

$$\int_{t_0}^{t_1} dt \cdot \sum_{1, 2, \cdots} \left(\frac{\partial H}{\delta\varphi_1} \delta\varphi_1 + \frac{\partial H}{\partial\dot{\varphi}_1} \delta\dot{\varphi}_1 + \Phi_1 \delta\varphi_1 \right) = 0.$$

From this follow the equations of motion:

$$\Phi_1 - \frac{d}{dt}\left(\frac{\partial H}{\partial\dot{\varphi}_1}\right) + \frac{\partial H}{\partial\varphi_1} = 0, \tag{59}$$

and so on for all the indices, 1, 2, $\cdots$. Through multiplication of the individual equations by $\dot{\varphi}_1, \dot{\varphi}_2, \cdots$ addition and integration with respect to time, there results the equation of conservation of energy, whereby the energy E is given by the expression:

$$E = \sum_{1, 2, \cdots} \dot{\varphi}_1 \frac{\partial H}{\partial\dot{\varphi}_1} - H. \tag{60}$$

In ordinary mechanics $H = L - U$, if L denote the kinetic and

U the potential energy. Since L is a homogeneous function of the second degree with respect to the $\dot{\varphi}$'s, it follows from (60) that:

$$E = 2L - H = L + U.$$

But this expression holds by no means in general.

We pass now to the consideration of the quasi-stationary motion of a system of linear conductors carrying simple closed galvanic currents. The state of the system is given by the position and the velocities of the conductors and by the current densities in each of the same. The coordinates referring to the position of the first conductor may be represented by $\varphi_1, \varphi_1', \varphi_1'', \cdots$, corresponding designations holding for the remaining conductors. We inquire now as to the increase of energy or the external work, A, which corresponds to a virtual displacement of all coordinates. Energy may be conveyed to the system through mechanical actions and through electromagnetic induction as well. The former corresponds to mechanical work, the latter to electromotive work. The former will be of the familiar form:

$$\Phi_1 \delta\varphi_1 + \Phi_1' \delta\varphi_1 + \cdots + \Phi_2 \delta\varphi_2 + \cdots.$$

If we denote by $E_1, E_2, \cdots$ the electromotive forces which are induced in the individual conductors through external agencies (e. g., moving magnets which do not belong to the system), then the electromotive work done from outside upon the currents in the conductors of the system is:

$$E_1 \delta\epsilon_1 + E_2 \delta\epsilon_2 + \cdots,$$

if $\delta\epsilon_1, \delta\epsilon_2, \cdots$ denote the quantities of electricity which pass through cross sections of the conductors due to infinitely small virtual currents. The finite current densities will then be denoted by $\dot{\epsilon}_1, \dot{\epsilon}_2, \cdots$. The electrical state of the first conductor is thus determined in general by the current density $\dot{\epsilon}_1$, the mechanical state (position and velocity) by the coordinates

$\varphi_1, \varphi_1', \varphi_1'', \cdots$ and the corresponding velocities $\dot\varphi_1, \dot\varphi_1', \dot\varphi_1'', \cdots$. The coordinates $\epsilon_1, \epsilon_2, \cdots$ are so-called "cyclical" coordinates, since the state does not depend upon their momentary values, but only upon their differential quotients with respect to time, just as, for example, the state of a body rotatable about an axis of symmetry depends only upon the angular velocity, and not upon the angle of rotation. The scheme of notation adopted permits of the direct application of the above formularization of the principle of least action to the case here considered. Thus $H = H_\phi + H_\epsilon$, where H_ϕ, the mechanical potential, depends only upon the φ's and $\dot\varphi$'s, while the electrokinetic potential H_ϵ takes the following form:

$$H_\epsilon = \tfrac{1}{2}L_{11}\dot\epsilon_1^2 + L_{12}\dot\epsilon_1\dot\epsilon_2 + L_{13}\dot\epsilon_1\dot\epsilon_3 + \cdots + \tfrac{1}{2}L_{22}\dot\epsilon_2^2 + \cdots.$$

The quantities $L_{11}, L_{12}, L_{13} \cdots L_{22}, \cdots$ the coefficients of self induction and mutual induction depend, however, in a definite manner upon the coordinates of position $\varphi_1, \varphi_1', \varphi_1'', \cdots, \varphi_2, \varphi_2', \varphi_2'', \cdots$.

In accordance with (59), we have for the motion of the first conductor:

$$\Phi_1 - \frac{d}{dt}\left(\frac{dH_\phi}{\partial\dot\varphi_1}\right) + \frac{\partial H_\phi}{\partial\varphi_1} + \frac{\partial H_\epsilon}{\partial\varphi_1} = 0,$$

with corresponding equations for $\varphi_1', \varphi_1'', \cdots$, and for the electric current in it:

$$E_1 - \frac{d}{dt}\left(\frac{\partial H_\epsilon}{\partial\dot\epsilon_1}\right) = 0.$$

The laws for the mechanical (ponderomotive) actions may be condensed into the statement that, in addition to the ordinary force upon the first conductor expressed by Φ_1, there is a mechanical force

$$\frac{\partial H_\epsilon}{\partial\varphi_1} = \frac{1}{2}\frac{\partial L_{11}}{\partial\varphi_1}\dot\epsilon_1^2 + \frac{\partial L_{12}}{\partial\varphi_1}\dot\epsilon_1\dot\epsilon_2 + \frac{\partial L_{13}}{\partial\varphi_1}\dot\epsilon_1\dot\epsilon_3 + \cdots,$$

which is composed of an action of the current upon itself (first term) and of the actions of the remaining currents upon it (following terms).

The laws of electrical action, on the other hand, are expressed by the statement, that to the external electromotive force E_1 in the first conductor there is added the electromotive force

$$-\frac{d}{dt}\left(\frac{dH_\epsilon}{\partial \dot{\epsilon}_1}\right) = -\frac{d}{dt}(L_{11}\dot{\epsilon}_1 + L_{12}\dot{\epsilon}_2 + L_{13}\dot{\epsilon}_3 + \cdots)$$

which likewise is composed of an action of the current upon itself (self induction) and of the inducing actions of the remaining currents, and that these two forces compensate each other.

The galvanic conductance or the galvanic resistance is not contained in these equations because the corresponding energy, Joule heat, is produced in an irreversible manner, and irreversible processes are not represented by the principle of least action. One can formally include this action, likewise any other irreversible action, in accordance with the procedure of Helmholz, by introducing it as an external force, in the present case as the electromotive force due to the resistance w, which operates to cause a diminution in the energy of the system. For an infinitely small element of time, the amount of this energy change is:

$$-(w_1\dot{\epsilon}_1^2 + w_2\dot{\epsilon}_2^2 + w_3\dot{\epsilon}_3^2 + \cdots)\cdot dt$$
$$= -(w_1\dot{\epsilon}_1 d\epsilon_1 + w_2\dot{\epsilon}_2 d\epsilon_2 + \cdots).$$

Consequently, since the external work $E_1 d\epsilon_1 + E_2 d\epsilon_2 + \cdots$ now includes the Joule heat, the external force components $E_1, E_2, \cdots$ in the electromotive equations must be increased by the additional terms $-w_1\dot{\epsilon}_1, -w_2\dot{\epsilon}_2, \cdots$.—

The application of the principle of least action to thermodynamic processes is of special interest, because the importance of the question relating to the fixing of the generalized coordinates, which determine the state of the system, here becomes prominent. From the standpoint of pure thermodynamics, the variables which determine the state of a body can certainly be quite arbitrarily chosen, e. g., in the case of a gas of invariable constitution any two of the following quantities may be chosen as independent variables and all others expressed through them: volume V, temperature T, pressure P, energy E, entropy S. In the present case, the matter is quite different. If we inquire, in order to apply the principle of least action, with regard to the energy change or the total work A which will be done upon the gas from without in an infinitely small virtual displacement, it may be written in the form:

$$A = -p \cdot \delta V + T \cdot \delta S.$$

$T\delta S$ is the heat added from without, $-p\delta V$ the mechanical work furnished from without. In order to bring this into agreement with the general formula for external work (58):

$$A = \Phi_1 \delta\varphi_1 + \Phi_2 \delta\varphi_2$$

it becomes necessary now to choose V and S as the generalized coordinates of state and, therefore, to identify with them the previously employed quantities φ_1 and φ_2. Then $-p$ and T are the generalized force components Φ_1 and Φ_2. Now, since in thermodynamics every reversible change of state proceeds with infinite slowness, the velocity components $\dot{V}$ and $\dot{S}$, and in general all differential coefficients with respect to time, are to be placed equal to zero, and the principle of least action (59) reduces to:

$$\Phi + \frac{\partial H}{\partial \varphi} = 0,$$

and, therefore, in our case:

$$-p + \left(\frac{\partial H}{\partial V}\right)_S = 0 \quad \text{and} \quad T + \left(\frac{\partial H}{\partial S}\right)_V = 0.$$

Further, in accordance with (60):

$$E = -H.$$

Now these equations are actually valid, since they only present other forms of the relation

$$dS = \frac{dE + pdV}{T}.$$

The view here presented is fundamentally that which is given in the energetics of Mach, Ostwald, Helm, and Wiedeburg. The generalized coordinates V and S are in this theory the "capacity factors," $-p$ and T the "intensity factors."[1] So long as one limits himself to an irreversible process, nothing stands in the way of carrying out this method completely, nor of a generalization to include chemical processes.

In opposition to it there is an essentially different method of regarding thermodynamic processes, which in its complete generality was first introduced into physics by Helmholtz. In accordance with this method, one generalized coordinate is V, and the other is not S, but a certain cyclical coordinate—we shall denote it, as in the previous example, by ϵ—which does not appear itself in the expression for the kinetic potential H and only appears through its differential coefficient, $\dot{\epsilon}$; and this differential coefficient is the temperature T. Accordingly, H is dependent only upon V and T. The equation for the total external work, in accordance with (58), is:

$$A = -p\delta V + E\delta\epsilon,$$

and agreement with thermodynamics is obviously found if we set:

$$E\delta\epsilon = T\delta S, \quad \text{and also:} \quad Ed\epsilon = TdS, \quad Edt = dS.$$

The equations (59) for the principle of least action become:

$$-p + \left(\frac{\partial H}{\partial V}\right)_T = 0 \quad \text{and} \quad E - \frac{d}{dt}\left(\frac{\partial H}{\partial T}\right)_V = 0,$$

or

$$d\left(\frac{\partial H}{\partial T}\right)_V = Edt = dS,$$

[1] The breaking up of the energy differentials into two factors by the exponents of energetics is by no means associated with a special property of energy, but is simply an expression for the elementary law that the differential of a function $F(x)$ is equal to the product of the differential dx by the derivative $\dot{F}(x)$.

or by integration:

$$\left(\frac{\partial H}{\partial T}\right)_V = S,$$

to an additive constant, which we may set equal to 0. For the energy there results, in accordance with (60):

$$E = \dot{\epsilon}\frac{\partial H}{\partial \dot{\epsilon}} - H = T\left(\frac{\partial H}{\partial T}\right)_V - H,$$

and consequently:

$$H = -(E - TS).$$

H is therefore equal to the negative of the function which Helmholz has called the "free energy" of the system, and the above equations are known from thermodynamics.

Furthermore, the method of Helmholz permits of being carried through consistently, and so long as one limits himself to the consideration of reversible processes, it is in general quite impossible to decide in favor of the one method or the other. However, the method of Helmholz possesses a distinct advantage over the other which I desire to emphasize here. It lends itself better to the furtherance of our endeavor toward the unification of the system of physics. In accordance with the purely energetic method, the independent variables V and S have absolutely nothing to do with each other; heat is a form of energy which is distinguished in nature from mechanical energy and which in no way can be referred back to it. In accordance with Helmholz, heat energy is reduced to motion, and this certainly indicates an advance which is to be placed, perhaps, upon exactly the same footing as the advance which is involved in the consideration of light waves as electromagnetic waves.

To be sure, the view of Helmholz is not broad enough to include irreversible processes; with regard to this, as we have earlier stated in detail, the introduction of the calculus of probability is necessary in order to throw light on the question. At the same time, this is also the real reason that the exponents of energetics will have nothing to do with the strict observance of irreversible processes, and they either declare them as doubtful or ignore them completely. In reality, the facts of the case are quite the reverse; irreversible processes are the only processes occurring in nature. Reversible processes form only an ideal abstraction, which is very valuable for the theory, but which is never completely realized in nature.

II. *The Generalized Coordinates of State Form a Continuous Manifold.*

The laws of infinitely small motions of perfectly elastic bodies furnish us with the simplest example. The coordinates of state are then the displacement components, $\mathfrak{v}_x$, $\mathfrak{v}_y$, $\mathfrak{v}_z$, of a material point from its position of equilibrium (x, y, z), considered as a function of the coordinates x, y, z. The external work is given by a surface integral:

$$A = \int d\sigma (X_\nu \delta \mathfrak{v}_x + Y_\nu \delta \mathfrak{v}_y + Z_\nu \delta \mathfrak{v}_z)$$

($d\sigma$, surface element; ν, inner normal). The kinetic potential is again given by the difference of the kinetic energy L and the potential energy U:

$$H = L - U.$$

The kinetic energy is:

$$L = \int \frac{d\tau k}{2} (\dot{\mathfrak{v}}_x^2 + \dot{\mathfrak{v}}_y^2 + \dot{\mathfrak{v}}_z^2),$$

wherein $d\tau$ denotes a volume element, k the volume density. The potential energy U is likewise a space integral of a homogeneous quadratic function f which specifies the potential energy of a volume element. This depends, as is seen from purely geometrical considerations, only upon the 6 "strain coefficients:"

$$\frac{\partial \mathfrak{v}_x}{\partial x} = x_x, \quad \frac{\partial \mathfrak{v}_y}{\partial y} = y_y, \quad \frac{\partial \mathfrak{v}_z}{\partial z} = z_z,$$

$$\frac{\partial \mathfrak{v}_y}{\partial z} + \frac{\partial \mathfrak{v}_z}{\partial y} = y_z = z_y, \quad \frac{\partial \mathfrak{v}_z}{\partial x} + \frac{\partial \mathfrak{v}_x}{\partial z} = z_x = x_z, \quad \frac{\partial \mathfrak{v}_x}{\partial y} + \frac{\partial \mathfrak{v}_y}{\partial x} = x_y = y_x.$$

In general, therefore, the function f contains 21 independent constants, which characterize the whole elastic behavior of the substance. For isotropic substances these reduce on grounds of symmetry to 2. Substituting these values in the expression for the principle of least action (57) we obtain:

$$\int dt \left\{ \int d\tau k(\dot{\mathfrak{v}}_x \delta \dot{\mathfrak{v}}_x + \cdots) - \int d\tau \left(\frac{\partial f}{\partial x_x} \delta x_x + \frac{\partial f}{\partial x_y} \delta x_y + \cdots \right) \right.$$
$$\left. + \int d\sigma (X_\nu \delta \mathfrak{v}_x + \cdots) \right\} = 0.$$

If we put for brevity:

$$-\frac{\partial f}{\partial x_x} = X_x, \qquad -\frac{\partial f}{\partial y_y} = Y_y, \qquad -\frac{\partial f}{\partial z_z} = Z_z,$$

$$-\frac{\partial f}{\partial y_z} = Y_z = Z_y, \quad -\frac{\partial f}{\partial z_x} = Z_x = X_z, \quad -\frac{\partial f}{\partial x_y} = X_y = Y_x,$$

it turns out, as the result of purely mathematical operations in which the variations $\delta \dot{\mathfrak{v}}_x, \delta \dot{\mathfrak{v}}_y, \cdots$ and likewise the variations $\delta x_x, \delta x_y, \cdots$ are reduced through suitable partial integration with respect to the variations $\delta \mathfrak{v}_x, \delta \mathfrak{v}_y, \cdots$, that the conditions within the body are expressed by:

$$k\ddot{\mathfrak{v}}_x + \frac{\partial X_x}{\partial x} + \frac{\partial X_y}{\partial y} + \frac{\partial X_z}{\partial z} = 0, \cdots$$

and at the surface, by:

$$X_\nu = X_x \cos \nu x + X_y \cos \nu y + X_z \cos \nu z, \cdots$$

as is known from the theory of elasticity. The mechanical significance of the quantities X_x, Y_y, $\cdots$ as surface forces follows from the surface conditions.

For the last application of the principle of least action we will take a special case of electrodynamics, namely, electrodynamic processes in a homogeneous isotropic non-conductor at rest, e. g., a vacuum. The treatment is analogous to that carried out in the foregoing example. The only difference lies in the fact that in electrodynamics the dependence of the potential energy U upon the generalized coordinate $\mathfrak{v}$ is somewhat different than in elastic phenomena.

We therefore again put for the external work:

$$A = \int d\sigma(X_\nu \delta\mathfrak{v}_x + Y_\nu \delta\mathfrak{v}_y + Z_\nu \delta\mathfrak{v}_z), \qquad (61)$$

and for the kinetic potential:

$$H = L - U,$$

wherein again:

$$L = \int d\tau \frac{k}{2}(\dot{\mathfrak{v}}_x^2 + \dot{\mathfrak{v}}_y^2 + \dot{\mathfrak{v}}_z^2) = \int d\tau \frac{k}{2}(\dot{\mathfrak{v}})^2.$$

On the other hand, we write here:

$$U = \int d\tau \frac{h}{2}(\text{curl } \mathfrak{v})^2.$$

Through these assumptions the dynamical equations including the boundary conditions are now completely determined. The principle of least action (57) furnishes:

$$\int dt \Big\{ \int d\tau k(\dot{\mathfrak{v}}_x \delta\dot{\mathfrak{v}}_x + \cdots) - \int d\tau h(\text{curl}_x \, \mathfrak{v}\delta \, \text{curl}_x \, \mathfrak{v} + \cdots) + \int d\sigma(X_\nu \delta\mathfrak{v}_x + \cdots) \Big\} = 0.$$

From this follow, in quite an analogous way to that employed above in the theory of elasticity, first, for the interior of the non-conductor:

$$k\ddot{\mathfrak{v}}_x = h\left(\frac{\partial \, \text{curl}_y \, \mathfrak{v}}{\partial z} - \frac{\partial \, \text{curl}_z \, \mathfrak{v}}{\partial y}\right), \quad \cdots$$

or more briefly

$$k\ddot{\mathfrak{v}} = - h \text{ curl curl } \mathfrak{v}, \qquad (62)$$

and secondly, for the surface:

$$X_\nu = h(\text{curl}_z \, \mathfrak{v} \cdot \cos \nu y - \text{curl}_y \, \mathfrak{v} \cdot \cos \nu z), \quad \cdots \qquad (63)$$

These equations are identical with the known electrodynamical equations, if we identify L with the electric, and U with the magnetic energy (or conversely). If we put

$$L = \frac{1}{8\pi}\int d\tau \cdot \epsilon\mathfrak{E}^2 \quad \text{and} \quad U = \frac{1}{8\pi}\int d\tau \cdot \mu\mathfrak{H}^2,$$

($\mathfrak{E}$ and $\mathfrak{H}$, the field strengths, ϵ, the dielectric constant, μ, the permeability) and compare these values with the above expressions for L and U we may write:

$$\dot{\mathfrak{v}} = -\mathfrak{E} \cdot \sqrt{\frac{\epsilon}{4\pi k}}, \quad \text{curl}\, \mathfrak{v} = \mathfrak{H}\sqrt{\frac{\mu}{4\pi h}}. \tag{64}$$

It follows then, by elimination of $\mathfrak{v}$, that:

$$\dot{\mathfrak{H}} = -\sqrt{\frac{\epsilon h}{\mu k}} \cdot \text{curl}\, \mathfrak{E},$$

and further, by substitution of $\dot{\mathfrak{v}}$ and curl $\mathfrak{v}$ in equation (62) found above for the interior of the non-conductor, that:

$$\dot{\mathfrak{E}} = \sqrt{\frac{\mu h}{\epsilon k}}\, \text{curl}\, \mathfrak{H}.$$

Comparison with the known electrodynamical equations expressed in Gaussian units:

$$\mu\dot{\mathfrak{H}} = -c\, \text{curl}\, \mathfrak{E}, \quad \epsilon\dot{\mathfrak{E}} = c\, \text{curl}\, \mathfrak{H}$$

(c, velocity of light in vacuum) results in a complete agreement, if we put:

$$\frac{c}{\mu} = \sqrt{\frac{\epsilon h}{\mu k}} \quad \text{and} \quad \frac{c}{\epsilon} = \sqrt{\frac{\mu h}{\epsilon k}}.$$

From either of these two equations it follows that:

$$\frac{h}{k} = \frac{c^2}{\epsilon\mu},$$

the square of the velocity of propagation.

We obtain from (61) for the energy entering the system from without:

$$dt \cdot \int d\sigma(X_\nu \dot{\mathfrak{v}}_x + Y_\nu \dot{\mathfrak{v}}_y + Z_\nu \dot{\mathfrak{v}}_z),$$

or, taking account of the surface equation (63):

$$dt \cdot \int d\sigma h\{(\text{curl}_z \mathfrak{v} \cos \nu y - \text{curl}_y \mathfrak{v} \cos \nu z)\dot{\mathfrak{v}}_x + \cdots\},$$

an expression which, upon substitution of the values of $\dot{\mathfrak{v}}$ and curl $\mathfrak{v}$ from (64), turns out to be identical with the Poynting energy current.

We have thus by an application of the principle of least action with a suitably chosen expression for the kinetic potential H arrived at the known Maxwellian field equations.

Are, then, the electromagnetic processes thus referred back to mechanical processes? By no means; for the vector $\mathfrak{v}$ employed here is certainly not a mechanical quantity. It is moreover not possible in general to interpret $\mathfrak{v}$ as a mechanical quantity, for instance, $\mathfrak{v}$ as a displacement, $\dot{\mathfrak{v}}$ as a velocity, curl $\mathfrak{v}$ as a rotation. Thus, e. g., in an electrostatic field $\dot{\mathfrak{v}}$ is constant. Therefore, $\mathfrak{v}$ increases with the time beyond all limits, and curl $\mathfrak{v}$ can no longer signify a rotation.[1] While from these considerations the possibility of a mechanical explanation of electrical phenomena is not proven, it does appear, on the other hand, to be undoubtedly true that the significance of the principle of least action may be essentially extended beyond ordinary mechanics and that this principle can therefore also be utilized as the foundation for general dynamics, since it governs all known reversible processes.

[1] With regard to the impossibility of interpreting electrodynamic processes in terms of the motions of a continuous medium, cf. particularly, H. Witte: Über den gegenwärtigen Stand der Frage nach einer mechanischen Erklärung der elektrischen Erscheinungen" Berlin, 1906 (E. Ebering).

EIGHTH LECTURE.

General Dynamics. Principle of Relativity.

In the lecture of yesterday we saw, by means of examples, that all continuous reversible processes of nature may be represented as consequences of the principle of least action, and that the whole course of such a process is uniquely determined as soon as we know, besides the actions which are exerted upon the system from without, the kinetic potential H as a function of the generalized coordinates and their differential coefficients with respect to time. The determination of this function remains then as a special problem, and we recognize here a rich field for further theories and hypotheses. It is my purpose to discuss with you today an hypothesis which represents a magnificent attempt to establish quite generally the dependency of the kinetic potential H upon the velocities, and which is commonly designated as the principle of relativity. The gist of this principle is: it is in no wise possible to detect the motion of a body relative to empty space; in fact, there is absolutely no physical sense in speaking of such a motion. If, therefore, two observers move with uniform but different velocities, then each of the two with exactly the same right may assert that with respect to empty space he is at rest, and there are no physical methods of measurement enabling us to decide in favor of the one or the other. The principle of relativity in its generalized form is a very recent development. The preparatory steps were taken by H. A. Lorentz, it was first generally formulated by A. Einstein, and was developed into a finished mathematical system by H. Minkowski. However, traces of it extend quite far back into the past, and therefore it seems desirable first to say something concerning the history of its development.

The principle of relativity has been recognized in mechanics since the time of Galilee and Newton. It is contained in the form of the simple equations of motion of a material point, since these contain only the acceleration and not the velocity of the point. If, therefore, we refer the motion of the point, first to the coordinates x, y, z, and again to the coordinates x', y', z' of a second system, whose axes are directed parallel to the first and which moves with the velocity ν in the direction of the positive x-axis:

$$x' = x - \nu t, \quad y' = y, \quad z' = z, \tag{65}$$

and the form of the equations of motion is not changed in the slightest. Nothing short of the assumption of the general validity of the relativity principle in mechanics can justify the inclusion by physics of the Copernican cosmical system, since through it the independence of all processes upon the earth of the progressive motion of the earth is secured. If one were obliged to take account of this motion, I should have, e. g., to admit that the piece of chalk in my hand possesses an enormous kinetic energy, corresponding to a velocity of something like 30 kilometers per second.

It was without doubt his conviction of the absolute validity of the principle of relativity which guided Heinrich Hertz in the establishment of his fundamental equations for the electrodynamics of moving bodies. The electrodynamics of Hertz is, in fact, wholly built upon the principle of relativity. It recognizes no absolute motion with regard to empty space. It speaks only of motions of material bodies relative to one another. In accordance with the theory of Hertz, all electrodynamic processes occur in material bodies; if these move, then the electrodynamic processes occurring therein move with them. To speak of an independent state of motion of a medium outside of material bodies, such as the ether, has just as little sense in the theory of Hertz as in the modern theory of relativity.

But the theory of Hertz has led to various contradictions with experience. I will refer here to the most important of these.

Fizeau brought (1851) into parallelism a bundle of rays originating in a light source L by means of a lens and then brought it to a focus by means of a second lens upon a screen S (Fig. 2).

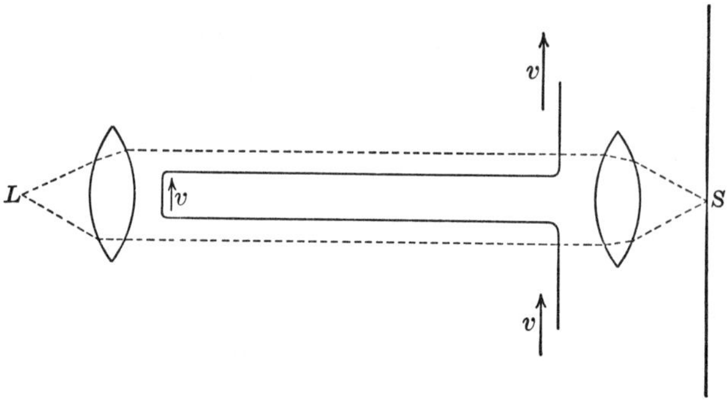

FIG. 2.

In the path of the parallel light rays between the two lenses he placed a tube system of such sort that a transparent liquid could be passed through it, and in such manner that in one half (the upper) the light rays would pass in the direction of flow of the liquid while in the other half (the lower), the rays would pass in the opposite direction.

If now a liquid or a gas flow through the tube system with the velocity ν, then, in accordance with the theory of Hertz, since light must be a process in the substance, the light waves must be transported with the velocity of the liquid. The velocity of light relative to L and S is, therefore, in the upper part $q_0 + \nu$, and the lower part $q_0 - \nu$, if q_0 denote the velocity of light relative to the liquid. The difference of these two velocities, 2ν, should be observable at S through corresponding interference of the lower and the upper light rays, and quite independently of the nature of the flowing substance. Experiment did not confirm this conclusion. Moreover, it showed in gases generally no trace of the expected action; i. e., light is propagated in a flowing gas in the same manner as in a gas at rest. On the other hand, in the case of liquids an effect was certainly indicated, but notably smaller in amount than that demanded by the theory of Hertz. Instead of the expected velocity difference $2v$, the difference $2v(1 - 1/n^2)$ only was observed, where n is the refractive index of the liquid. The factor $(1 - 1/n^2)$ is called the Fresnel coefficient. There is contained (for $n = 1$) in this expression the result obtained in the case of gases.

It follows from the experiment of Fizeau that, as regards electrodynamic processes in a gas, the motion of the gas is practically immaterial. If, therefore, one holds that electrodynamic processes require for their propagation a substantial carrier, a special medium, then it must be concluded that this medium, the ether, remains at rest when the gas moves in an arbitrary manner. This interpretation forms the basis of the electrodynamics of Lorentz, involving an absolutely quiescent ether. In accordance with this theory, electrodynamic phenomena have only indirectly to do with the motion of matter. Primarily all electrodynamical actions are propagated in ether at rest. Matter influences the propagation only in a secondary way, so far as it is the cause of exciting in greater or less degree resonant vibrations in its smallest parts by means of the electrodynamic waves passing through it. Now, since the refractive properties of substances are also influenced through the resonant vibrations of its smallest particles, there results from this theory a definite connection between the refractive index and the coefficient of Fresnel, and this connection is, as calculation shows, exactly that demanded by measurements. So far, therefore, the theory of Lorentz is confirmed through experience, and the principle of relativity is divested of its general significance.

The principle of relativity was immediately confronted by a new difficulty. The theory of a quiescent ether admits the idea of an absolute velocity of a body, namely the velocity relative to the ether. Therefore, in accordance with this theory, of two observers A and B who are in empty space and who move relatively to each other with the uniform velocity v, it would be at best possible for only one rightly to assert that he is at rest relative to the ether. If we assume, e. g., that at the moment at which the two observers meet an instantaneous optical signal, a flash, is made by each, then an infinitely thin spherical wave spreads out from the place of its origin in all directions through empty space. If, therefore, the observer A remain at the center of the sphere, the observer B will not remain at the center and, as judged by the observer B, the light in his own direction of motion must travel (with the velocity $c - \nu$) more slowly than in the opposite direction (with the velocity $c + \nu$), or than in a perpendicular direction (with the velocity $\sqrt{c^2 - \nu^2}$) (cf. Fig. 3).

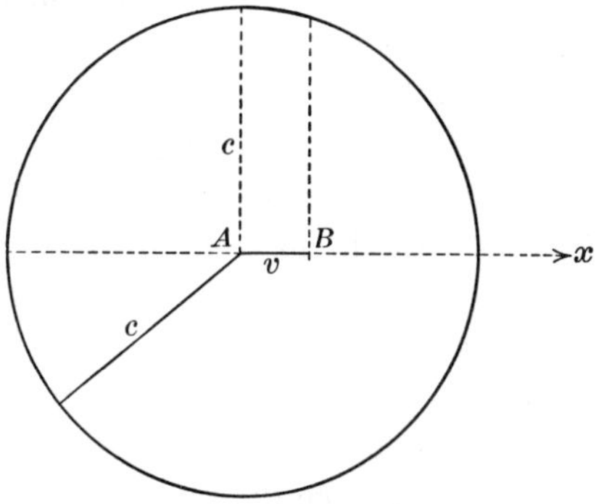

Fig. 3.

Under suitable conditions the observer B should be able to detect and measure this sort of effect.

This elementary consideration led to the celebrated attempt of Michelson to measure the motion of the earth relative to the ether. A parallel beam of rays proceeding from L (Fig. 4) falls upon a transparent plane parallel plate P inclined at 45°, by which it is in part transmitted and in part reflected. The transmitted and reflected beams are brought into interference by reflection from suitable metallic mirrors S_1 and S_2, which are removed by the same distance l from P. If, now, the earth with the whole apparatus moves in the direction PS_1 with the velocity ν, then the time which the light needs in order to go from P to S_1 and back is:

$$\frac{l}{c-v}+\frac{l}{c+v}=\frac{2l}{c}\left(1+\frac{v^2}{c^2}+\cdots\right).$$

On the other hand, the time which the light needs in order to pass from P to S_2 and back to P is:

$$\frac{l}{\sqrt{c^2-v^2}}+\frac{l}{\sqrt{c^2-v^2}}=\frac{2l}{c}\left(1+\frac{1}{2}\frac{v^2}{c^2}+\cdots\right).$$

If, now, the whole apparatus be turned through a right angle, a noticeable displacement of the interference bands should result,

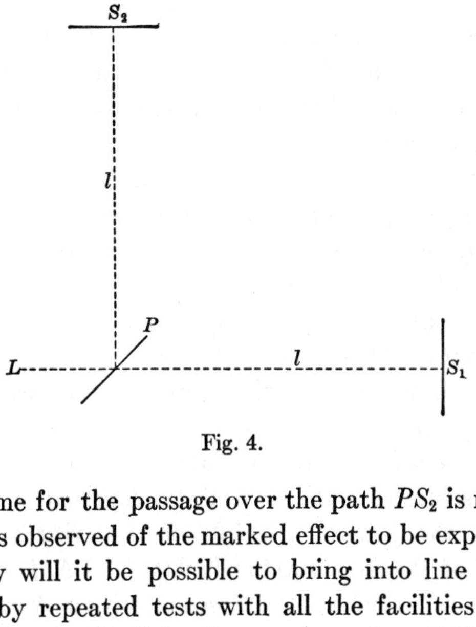

Fig. 4.

since the time for the passage over the path PS_2 is now longer. No trace was observed of the marked effect to be expected.

Now, how will it be possible to bring into line this result, established by repeated tests with all the facilities of modern experimental art? E. Cohn has attempted to find the necessary compensation in a certain influence of the air in which the rays are propagated. But for anyone who bears in mind the great results of the atomic theory of dispersion and who does not renounce the simple explanation which this theory gives for the dependence of the refractive index upon the color, without introducing something else in its place, the idea that a moving absolutely transparent medium, whose refractive index is absolutely = 1, shall yet have a notable influence upon the velocity of propagation of light, as the theory of Cohn demands, is not possible of assumption. For this theory distinguishes essentially a transparent medium, whose refractive index is = 1, from a perfect vacuum. For the former the velocity of propagation of light in the direction of the velocity ν of the medium with relation to an observer at rest is

$$q = c + \frac{\nu^2}{c},$$

for a vacuum, on the other hand, $q = c$. In the former medium, Cohn's theory of the Michelson experiment predicts no effect, but, on the other hand, the Michelson experiment should give a positive effect in a vacuum.

In opposition to E. Cohn, H. A. Lorentz and FitzGerald ascribe the necessary compensation to a contraction of the whole optical apparatus in the direction of the earth's motion of the order of magnitude ν^2/c^2. This assumption allows better of the introduction again of the principle of relativity, but it can first completely satisfy this principle when it appears, not as a necessary hypothesis made to fit the present special case, but as a consequence of a much more general postulate. We have to thank A. Einstein for the framing of this postulate and H. Minkowski for its further mathematical development.

Above all, the general principle of relativity demands the renunciation of the assumption which led H. A. Lorentz to the framing of his theory of a quiescent ether; the assumption of a substantial carrier of electromagnetic waves. For, when such a carrier is present, one must assume a definite velocity of a ponderable body as definable with respect to it, and this is exactly that which is excluded by the relativity principle. Thus the ether drops out of the theory and with it the possibility of mechanical explanation of electrodynamic processes, i. e., of referring them to motions. The latter difficulty, however, does not signify here so much, since it was already known before, that no mechanical theory founded upon the continuous motions of the ether permits of being completely carried through (cf. p. 111). In place of the so-called free ether there is now substituted the absolute vacuum, in which electromagnetic energy is independently propagated, like ponderable atoms. I believe it follows as a consequence that no physical properties can be consistently ascribed to the absolute vacuum. The dielectric constant and the magnetic permeability of a vacuum have no absolute meaning, only relative. If an electrodynamic process were to occur in a ponderable medium as in a vacuum, then it would have absolutely no sense to distinguish between field strength and induction. In fact, one can ascribe to the vacuum any arbitrary value of the dielectric constant, as is indicated by the various systems of units. But how is it now with regard to the velocity of propagation of light? This also is not to be regarded as a property of the vacuum, but as a property of electromagnetic energy which is present in the vacuum. Where there is no energy there can exist no velocity of propagation.

With the complete elimination of the ether, the opportunity is now present for the framing of the principle of relativity. Obviously, we must, as a simple consideration shows, introduce something radically new. In order that the moving observer B mentioned above (Fig. 3, p. 116) shall not see the light signal given by him travelling more slowly in his own direction of motion (with the velocity $c - \nu$) than in the opposite direction (with the velocity $c + \nu$), it is necessary that he shall not identify the instant of time at which the light has covered the distance $c - \nu$ in the direction of his own motion with the instant of time at which the light has covered the distance $c + \nu$ in the opposite direction, but that he regard the latter instant of time as later. In other words: the observer B measures time differently from the observer A. This is a priori quite permissible; for the relativity principle only demands that neither of the two observers shall come into contradiction with himself. However, the possibility is left open that the specifications of time of both observers may be mutually contradictory.

It need scarcely be emphasized that this new conception of the idea of time makes the most serious demands upon the capacity of abstraction and the projective power of the physicist. It surpasses in boldness everything previously suggested in speculative natural phenomena and even in the philosophical theories of knowledge: non-euclidean geometry is child's play in comparison. And, moreover, the principle of relativity, unlike non-euclidean geometry, which only comes seriously into consideration in pure mathematics, undoubtedly possesses a real physical significance. The revolution introduced by this principle into the physical conceptions of the world is only to be compared in extent and depth with that brought about by the introduction of the Copernican system of the universe.

Since it is difficult, on account of our habitual notions concerning the idea of absolute time, to protect ourselves, without special carefully considered rules, against logical mistakes in the necessary processes of thought, we shall adopt the mathematical method of treatment. Let us consider then an electrodynamic process in a pure vacuum; first, from the standpoint of an observer A; secondly, from the standpoint of an observer B, who moves relatively to observer A with a velocity ν in the direction of the x-axis. Then, if A employ the system of reference x, y, z, t, and B the system of reference x', y', z', t', our first problem is to find the relations among the primed and the unprimed quantities. Above all, it is to be noticed that since both systems of reference, the primed and the unprimed, are to be like directed, the equations of transformation between corresponding quantities in the two systems must be so established that it is possible through a transformation of exactly the same kind to pass from the first system to the second, and conversely, from the second back to the first system. It follows immediately from this that the velocity of light c' in a vacuum for the observer B is exactly the same as for the observer A. Thus, if c' and c are different, $c' > c$, say, it would follow that: if one passes from one observer A to another observer B who moves with respect to A with uniform velocity, then he would find the velocity of propagation of light for B greater than for A. This conclusion must likewise hold quite in general independently of the direction in which B moves with respect to A, because all directions in space are equivalent for the observer A. On the same grounds, in passing from B to A, c must be greater than c', for all directions in space for the observer B are now equivalent. Since the two inequalities contradict, therefore c' must be equal to c. Of course this important result may be generalized immediately, so that the totality of the quantities independent of the motion, such as the velocity of light in a vacuum, the constant of gravitation between two bodies at rest, every isolated electric charge, and the entropy of any physical system possess the same values for both observers. On the other hand, this law does not hold for quantities such as energy, volume, temperature, etc. For these quantities depend also upon the velocity, and a body which is at rest for A is for B a moving body.

We inquire now with regard to the form of the equations of transformation between the unprimed and the primed coordinates. For this purpose let us consider, returning to the previous example, the propagation, as it appears to the two observers A and B, of an instantaneous signal creating an infinitely thin light wave which, at the instant at which the observers meet, begins to spread out from the common origin of coordinates. For the observer A the wave travels out as a spherical wave:

$$x^2 + y^2 + z^2 - c^2t^2 = 0. \tag{66}$$

For the second observer B the same wave also travels as a spherical wave with the same velocity:

$$x'^2 + y'^2 + z'^2 - c^2t'^2 = 0; \tag{67}$$

for the first observer has no advantage over the second observer.

B can exactly with the same right as *A* assert that he is at rest at the center of the spherical wave, and for *B*, after unit time, the

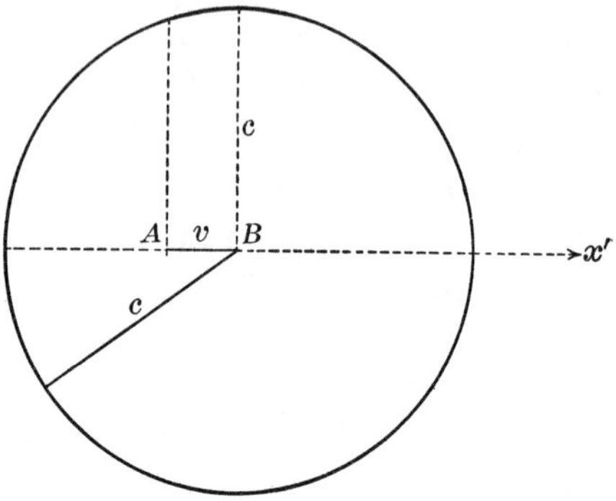

Fig. 5.

wave appears as in Fig. 5, while its appearance for the observer *A* after unit time, is represented by Fig. 3 (p. 116).[1]

The equations of transformation must therefore fulfill the condition that the two last equations, which represent the same physical process, are compatible with each other; and furthermore: the passage from the unprimed to the primed quantities must in no wise be distinguished from the reverse passage from the primed to the unprimed quantities. In order to satisfy these conditions, we generalize the equations of transformation (65), set up at the beginning of this lecture for the old mechanical principle of relativity, in the following manner:

$$x' = \kappa(x - vt), \quad y' = \lambda y, \quad z' = \mu z, \quad t' = \nu t + \rho x.$$

Here v denotes, as formerly, the velocity of the observer *B* relative to *A* and the constants $\kappa, \lambda, \mu, \nu, \rho$ are yet to be determined. We must have:

$$x = \kappa'(x' - v't'), \quad y = \lambda' y', \quad z = \mu' z', \quad t = \nu' t' + \rho' x'.$$

It is now easy to see that λ and λ' must both $= 1$. For, if, e. g.,

[1] The circumstance that the signal is a finite one, however small the time may be, has significance only as regards the thickness of the spherical layer and not for the conclusions here under consideration.

λ be greater than 1, then λ' must also be greater than 1; for the two transformations are equivalent with regard to the y axis. In particular, it is impossible that λ and λ' depend upon the direction of motion of the other observer. But now, since, in accordance with what precedes, $\lambda = 1/\lambda'$, each of the two inequalities contradict and therefore $\lambda = \lambda' = 1$; likewise, $\mu = \mu' = 1$. The condition for identity of the two spherical waves then demands that the expression (66):

$$x^2 + y^2 + z^2 - c^2t^2$$

become, through the transformation of coordinates, identical with the expression (67):

$$x'^2 + y'^2 + z'^2 - c^2t'^2,$$

and from this the equations of transformation follow without ambiguity:

$$x' = \kappa(x - \nu t), \quad y' = y, \quad z' = z, \quad t' = \kappa\left(t - \frac{\nu}{c^2}x\right), \qquad (68)$$

wherein

$$\kappa = \frac{c}{\sqrt{c^2 - \nu^2}}.$$

Conversely:

$$x = \kappa(x' + vt'), \quad y = y', \quad z = z', \quad t = \kappa\left(t' + \frac{\nu}{c^2}x'\right). \qquad (69)$$

These equations permit quite in general of the passage from the system of reference of one observer to that of the other (H. A. Lorentz), and the principle of relativity asserts that all processes in nature occur in accordance with the same laws and with the same constants for both observers (A. Einstein). Mathematically considered, the equations of transformation correspond to a rotation in the four dimensional system of reference (x, y, z, ict) through the imaginary angle arctg ($i(\nu/c)$) (H. Minkowski). Accordingly, the principle of relativity simply teaches that there is in the four dimensional system of space and time no special characteristic direction, and any doubts concerning the general validity of the principle are of exactly the same kind as those concerning the existence of the antipodians upon the other side of the earth.

We will first make some applications of the principle of relativity to processes which we have already treated above. That the result of the Michelson experiment is in agreement with the principle of relativity, is immediately evident; for, in accordance with the relativity principle, the influence of a uniform motion of the earth upon processes on the earth can under no conditions be detected.

We consider now the Fizeau experiment with the flowing liquid (see p. 114). If the velocity of propagation of light in the liquid at rest be again q_0, then, in accordance with the relativity principle, q_0 is also the velocity of the propagation of light in the flowing liquid for an observer who moves with the liquid, in case we disregard the dispersion of the liquid; for the color of the light is different for the moving observer. If we call this observer B and the velocity of the liquid as above, ν, we may employ immediately the above formulae in the calculation of the velocity of propagation of light in the flowing liquid, judged by an observer A at the screen S. We have only to put

$$\frac{dx'}{dt'} = \dot{x}' = q_0,$$

and to seek the corresponding value of

$$\frac{dx}{dt} = \dot{x}.$$

For this obviously gives the velocity sought.

Now it follows directly from the equations of transformation (69) that:

$$\frac{dx}{dt} = \dot{x} = \frac{\dot{x}' + \nu}{1 + \frac{\nu \dot{x}'}{c^2}},$$

and, therefore, through appropriate substitution, the velocity sought in the upper tube, after neglecting higher powers in ν/c and ν/q_0, is:

$$\dot{x} = \frac{q_0 + \nu}{1 + \frac{\nu q_0}{c^2}} = q_0 + \nu\left(1 - \frac{q_0^2}{c^2}\right),$$

and the corresponding velocity in the lower tube is:

$$q_0 - \nu\left(1 - \frac{q_0^2}{c^2}\right).$$

The difference of the two velocities is

$$2\nu\left(1 - \frac{q_0^2}{c^2}\right) = 2\nu\left(1 - \frac{1}{n^2}\right),$$

which is the Fresnel coefficient, in agreement with the measurements of Fizeau.

The significance of the principle of relativity extends, not only to optical and other electrodynamic phenomena, but also to all processes of ordinary mechanics; but the familiar expression ($\frac{1}{2}mq^2$) for the kinetic energy of a mass point moving with the velocity q is incompatible with this principle.

But, on the other hand, since all mechanics as well as the rest of physics is governed by the principle of least action, the significance of the relativity principle extends at bottom only to the particular form which it prescribes for the kinetic potential H, and this form, though I will not stop to prove it, is characterized by the simple law that the expression

$$H \cdot dt$$

for every space element of a physical system is an invariant

$$= H' \cdot dt'$$

with respect to the passage from one observer A to the other observer B or, what is the same thing, the expression $H/\sqrt{c^2 - q^2}$ is in this passage an invariant $= H'/\sqrt{c^2 - q'^2}$.

Let us now make some applications of this very general law, first to the dynamics of a single mass point in a vacuum, whose state is determined by its velocity q. Let us call the kinetic potential of the mass point for $q = 0$, H_0, and consider now the point at an instant when its velocity is q. For an observer B who moves with the velocity q with respect to the observer A, $q' = 0$ at this instant, and therefore $H' = H_0$. But now since in general:

$$\frac{H}{\sqrt{c^2 - q^2}} = \frac{H'}{\sqrt{c^2 - q'^2}},$$

we have after substitution:

$$H = \sqrt{1 - \frac{q^2}{c^2}} \cdot H_0 = \sqrt{1 - \frac{\dot{x}^2 + \dot{y}^2 + \dot{z}^2}{c^2}} \cdot H_0.$$

With this value of H, the Lagrangian equations of motion (59) of the previous lecture are applicable.

In accordance with (60), the kinetic energy of the mass point amounts to:

$$E = \dot{x}\frac{\partial H}{\partial \dot{x}} + \dot{y}\frac{\partial H}{\partial \dot{y}} + \dot{z}\frac{\partial H}{\partial \dot{z}} - H = q\frac{\partial H}{\partial q} - H = -\frac{H_0}{\sqrt{1 - \frac{q^2}{c^2}}},$$

and the momentum to:

$$G = \frac{\partial H}{\partial q} = -\frac{qH_0}{c\sqrt{c^2 - q^2}}.$$

G/q is called the transverse mass m_t, and dG/dq the longitudinal mass m_l of the point; accordingly:

$$m_t = -\frac{H_0}{c\sqrt{c^2 - q^2}}, \quad m_l = -\frac{cH_0}{(c^2 - q^2)^{3/2}}.$$

For $q = 0$, we have

$$m_t = m_l = m_0 = -\frac{H_0}{c^2}.$$

It is apparent, if one replaces in the above expressions the constant H_0 by the constant m_0, that the momentum is:

$$G = \frac{m_0 q}{\sqrt{1 - \frac{q^2}{c^2}}}$$

and the transverse mass:

$$m_t = \frac{m_0}{\sqrt{1 - \frac{q^2}{c^2}}},$$

and the longitudinal mass:

$$m_l = \frac{m_0}{\left(1 - \frac{q^2}{c^2}\right)^{3/2}},$$

and, finally, that the kinetic energy is:

$$E = \frac{m_0 c^2}{\sqrt{1 - \frac{q^2}{c^2}}} = m_0 c^2 + \tfrac{1}{2} m_0 q^2 + \cdots.$$

The familiar value of ordinary mechanics $\frac{1}{2} m_0 q^2$ appears here therefore only as an approximate value. These equations have been experimentally tested and confirmed through the measurements of A. H. Bucherer and of E. Hupka upon the magnetic deflection of electrons.

A further example of the invariance of $H \cdot dt$ will be taken from electrodynamics. Let us consider in any given medium any electromagnetic field. For any volume element V of the medium, the law holds that $V \cdot dt$ is invariant in the passage from the one to the other observer. It follows from this that H/V is invariant;

i. e., the kinetic potential of a unit volume or the "*space density of kinetic potential*" is invariant.

Hence the following relation exists;

$$\mathfrak{E}\mathfrak{D} - \mathfrak{H}\mathfrak{B} = \mathfrak{E}'\mathfrak{D}' - \mathfrak{H}'\mathfrak{B}',$$

wherein $\mathfrak{E}$ and $\mathfrak{H}$ denote the field strengths and $\mathfrak{D}$ and $\mathfrak{B}$ the corresponding inductions. Obviously a corresponding law for the space energy density $\mathfrak{E}\mathfrak{D} + \mathfrak{H}\mathfrak{B}$ will not hold.

A third example is selected from thermodynamics. If we take the velocity q of a moving body, the volume V and the temperature T as independent variables, then, as I have shown in the previous lecture (p. 105), we shall have for the pressure p and the entropy S the following relations:

$$\frac{\partial H}{\partial V} = p \quad \text{and} \quad \frac{\partial H}{\partial T} = S.$$

Now since $V/\sqrt{c^2 - q^2}$ is invariant, and S likewise invariant (see p. 121), it follows from the invariance of $H/\sqrt{c^2 - q^2}$ that p is invariant and also that $T/\sqrt{c^2 - q^2}$ is invariant, and hence that:

$$p = p' \quad \text{and} \quad \frac{T}{\sqrt{c^2 - q^2}} = \frac{T'}{\sqrt{c^2 - q'^2}}.$$

The two observers A and B would estimate the pressure of a body as the same, but the temperature of the body as different.

A special case of this example is supplied when the body considered furnishes a black body radiation. The black body radiation is the only physical system whose dynamics (for quasi-stationary processes) is known with absolute accuracy. That the black body radiation possesses inertia was first pointed out by F. Hasenöhrl. For black body radiation at rest the energy $E_0 = aT^4V$ is given by the Stefan-Boltzmann law, and the entropy $S_0 = \int (dE_0/T) = \frac{4}{3}aT^3V$, and the pressure $p_0 = (a/3)T^4$, and, therefore, in accordance with the above relations, the kinetic potential is:

$$H_0 = \frac{a}{3} T^4 V.$$

Let us imagine now a black body radiation moving with the velocity q with respect to the observer A and introduce an observer B who is at rest ($q = 0$) with reference to the black body radiation; then:

$$\frac{H}{\sqrt{c^2 - q^2}} = \frac{H'}{\sqrt{c^2 - q'^2}} = \frac{H_0'}{c},$$

wherein

$$H_0' = \frac{a}{3} T'^4 V'.$$

Taking account of the above general relations between T' and T, V' and V, this gives for the moving black body radiation the kinetic potential:

$$H = \frac{a}{3} \frac{T^4 V}{\left(1 - \frac{q^2}{c^2}\right)^2},$$

from which all the remaining thermodynamic quantities: the pressure p, the energy E, the momentum G, the longitudinal and transverse masses m_l and m_t of the moving black body radiation are uniquely determined.—

Colleagues, ladies and gentlemen, I have arrived at the conclusion of my lectures. I have endeavored to bring before you in bold outline those characteristic advances in the present system of physics which in my opinion are the most important. Another in my place would perhaps have made another and better choice and, at another time, it is quite likely that I myself should have done so. The principle of relativity holds, not only for processes in physics, but also for the physicist himself, in that a fixed system of physics exists in reality only for a given physicist and for a given time. But, as in the theory of relativity, there exist invariants in the system of physics: ideas and laws which retain their meaning for all investigators and for all times, and to discover these invariants is always the real endeavor of physical research. We shall work further in this direction in order to leave behind for our successors where possible—lasting results. For if, while engaged in body and mind in patient and often modest individual endeavor, one thought strengthens and supports us, it is this, that we in physics work, not for the day only and for immediate results, but, so to speak, for eternity.

I thank you heartily for the encouragement which you have given me. I thank you no less for the patience with which you have followed my lectures to the end, and I trust that it may be possible for many among you to furnish in the direction indicated much valuable service to our beloved science.

Notes on Planck's *Lectures on Theoretical Physics*

[Line numbers are counted from the top of Planck's pages, except those marked *b*, which are counted from the bottom.]

FIRST LECTURE

p. 2, l. 9 [Planck's teacher at Munich, Philipp von Jolly, advised him against a career in physics on the ground that the discovery of the principles of thermodynamics had completed the structure of theoretical physics. See J. L. Heilbron, *Dilemmas*, p. 10 and Lawrence Badash, "Completeness of Nineteenth Century Science," *Isis*, **63**, 48–58 (1972).]

p. 4, l. 5 [Planck here refers implicitly to Ernst Mach, wishing "to avoid anything that would personally offend the worthy old gentleman, but I must for once stand up to his 'antimetaphysical' theory. To that I am obliged by my own convictions" (letter to Laue, August 5, 1910). See Erwin N. Hiebert, *The Concept of Thermodynamics in the Scientific Thought of Mach and Planck* (Freiburg: Ernst-Mach-Institut, 1968), J. T. Blackmore, *Ernst Mach* (Berkeley: University of California Press, 1972), pp. 217–222, and Needell, "Irreversibility," pp. 108–110.]

p. 8, l. 4b [Lorentz, and Paul Drude before him, argued that electrons in a metal might be considered as a kind of "gas" because they are essentially free (since weakly interacting). Applying the kinetic theory of gases then yields experimental predictions. For his Columbia lectures of 1906 Lorentz produced his most extensive exposition of this work, *The Theory of Electrons* (Dover, 1952), which refers briefly to Planck's work on pp. 78–80 and reviews Drude's work on pp. 63–67.]

p. 9, l. 7b [The classic papers by Robert Mayer ("The Forces of Inorganic Nature," 1842), James Joule ("On Matter, Living Force, and Heat," 1847), and Hermann von Helmholtz ("The Conservation of Force," 1847) are included in *Kinetic Theory*, ed. Stephen G. Brush (Oxford: Pergamon Press, 1965), vol. 1, pp. 71–110.]

p. 10, l. 19 [Clausius formulated the first and second laws of thermodynamics as: "1. The energy of the universe is constant. 2. The entropy of the universe strives to attain a maximum value." See his 1865 paper "On Different Forms of the Fundamental Equations of the Mechanical Theory of Heat," reprinted in *The Second Law of Thermodynamics*, ed. Joseph Kestin (Stroudsburg, PA: Dowden, Hutchinson, & Ross, 1976), pp. 162–193, at 193.]

p. 11, l. 5 [Wilhelm Ostwald, Georg Helm, and others put forward a concept of "energetics" in which the second law of thermodynamics is not merely a statistical measure of the likeliness of states of increasing disorder, but rather a deterministic law which does not require an underlying atomic theory; when he began his work on black-body radiation, Planck was close to this opinion. See Stephen G. Brush, *The Kind of Motion We Call Heat* (Amsterdam: North-Holland, 1976), vol. 1, pp. 61–62, 96–98 and Erwin N. Hiebert, "The Energetics Controversy and the New Thermodynamics" in *Perspectives in the History of Science and Technology*, ed. Duane H. D. Roller (Norman, OK: University of Oklahoma Press, 1971), pp. 67–97; Ostwald's 1895 lecture "The Failure of Scientific Materialism" is translated in *Popular Science Monthly* **48**, 589–601 (March 1896).]

p. 13, l. 11b [This is the formulation of the second law which Planck had introduced in his doctoral thesis (1878); for a more extensive exposition see his *Treatise on Thermodynamics*, pp. 89–107.]

p. 15, l. 1 [The "incorrect conception" was that Carnot considered heat to be an actual material substance, called "caloric." This conception was abandoned by most physicists by 1840; see Brush, *Heat*, vol. 1, pp. 27–32. Interestingly, Carnot's analysis did not depend on this assumption and survived the abandonment of the caloric theory; see Brush, *Heat*, vol. 2, pp. 566–583 and Martin J. Klein, "Carnot's contribution to thermodynamics," *Physics Today* **27**:8, 23–28 (1974).]

p. 15, l. 8b [For Carnot's original demonstration (1824) that there is a greatest possible amount of work that can be extracted from an idealized heat engine, see his *Reflections on the Motive Power of Fire* (Dover, 1960), pp. 3–22.]

p. 16, l. 8 [Planck is thinking of heat transferred from a warmer body to a colder one. But depending on whether we identify the heat transferred with (1) the amount given up or (2) the amount absorbed, we get two different expressions for the work that has been lost. Now the work that can be obtained from an ideal Carnot engine depends on the difference of the two temperatures, T_1-T_2. The work done also is proportional to the quantity of heat transferred, Q. But depending on whether we are considering case (1) or case (2) we will form the ratio of $Q(T_1-T_2)$ with either (1) T_1 or (2) T_2. Then the work lost is either (1) $Q(T_1-T_2)/T_1$ or (2) $Q(T_1-T_2)/T_2$, and Planck notes that the mathematical form of the result is not unique.]

p. 17, l. 6b [Clausius does not specify the reversible process; he only postulates that such a reversible process exists in order to make his definition of entropy consistent; for example, a part of a Carnot cycle will satisfy the requirement of reversibility. Since Clausius' entropy is a function only of the *state* of the system, and not of how it got there, he argues that it can characterize even irreversible processes.]

p. 18, l. 4b [Planck chooses to speak of heat *loss*, which makes his Q negative here. Hence the last inequality on the page might be written $\Sigma\ (-Q)/T \geqslant 0$, showing that the magnitude of the heat expended ($-Q$) is positive and thus the entropy must increase.]

p. 19, l. 1 [*Isothermal* means that the process occurs at constant temperature. The two bottom equations on p. 18 then imply that $\Sigma Q \leqslant 0$ and that $A \leqslant 0$.]

p. 19, l. 4b [See the Seventh Lecture; Newton's laws and Maxwell's equations are reversible in time and thus can be formulated in terms of a principle of least action. Planck also treats these matters in "The Principle of Least Action," *Survey of Physical Theory*, pp. 69–81 (*PAV* 3.91–101). For Helmholtz's treatment see Wolfgang Yourgrau and Stanley Mandelstam, *Variational Principles in Dynamics and Quantum Theory* (Dover, 1979), pp. 93–95, 163–164.]

SECOND LECTURE

p. 21, l. 17 [Planck takes examples from his own work from 1878–1895, which he treated in more detail in his *Treatise on Thermodynamics*, pp. 139–292; his original papers (*PAV* 1.1-444) are discussed by Kangro, *Early History*, pp. 110–124. Although he did not know it at the time, the same results had been found by the American physicist Josiah Willard Gibbs (1839–1903); see Martin J. Klein, "The Scientific Style of Josiah Willard Gibbs," in *Springs of Scientific Creativity*, ed. Rutherford Aris *et al.* (Minneapolis: University of Minnesota Press, 1983), pp. 142–162. After reading Gibbs's *Elementary Principles of Statistical Mechanics* (1902; Dover, 1960) Planck wrote a review which expressed both admiration as well as criticism, particularly concerning Gibbs's treatment of "the question of the identity of the particles"; see Needell, "Irreversibility," pp. 45–55, Pesic, "Identicality," and Paul and Tatiana Ehrenfest, *The Conceptual Foundations of the Statistical Approach in Mechanics*, tr. Michael J. Moravcsik (Ithaca: Cornell University Press, 1959), p. 62. By 1909, though, Planck increasingly appreciated what Gibbs had accomplished, as his tribute here records.]

p. 24, l. 3 [Note that lower-case $u_1, u_2, \ldots$ and $v_1, v_2, \ldots$ are the energy and volume per molecule for each phase; u_0 and v_0 refer to the solvent.]

p. 25, l. 3b [*Isobaric* means the process occurs at constant pressure.]

p. 31, l. 3 [Clapeyron had derived this equation to relate the relative ratios of heat loss and slope of pressure/temperature lines predicted by Carnot's treatment of ideal engines; see Clapeyron's "Memoir on the Motive Power of Heat," in *On the Motive Power of Heat*, pp. 73–105 and Brush, *Heat*, vol. 2, pp. 566–583.]

THIRD LECTURE

p. 44, l. 4b [Note the important hypothesis here: the probability of a certain physical state can be expressed as the sum of the probabilities of the molecular configurations or "complexions" which can go up to make up that state; those complexions are considered to have equal *a priori* probabilities. This approach Planck will apply at a crucial moment of his black body radiation calculation (pp. 90–91). For Boltzmann's use of this technique see Martin J. Klein, "The Development of Boltzmann's Statistical Ideas" in *The Boltzmann Equation*, ed. E. G. D. Cohen and W. Thirring (New York: Springer-Verlag, 1973), pp. 53–106 and "Max Planck and the Beginnings of Quantum Theory," pp. 472–476; see also Kuhn, *Black-Body Theory*, pp. 57–60, 98, 101, 105, Olivier Darrigol, "Statistics and combinatorics in early quantum theory," *Historical Studies in the Physical Sciences* **19**, 17–80 (1988), Ulrich Hoyer, "Von Boltzmann zu Planck," *Archive for History of Exact Sciences* **23**, 47–86 (1980), and Martin Koch, "From Boltzmann to Planck: On Continuity in Scientific Revolutions" in *World Views and Scientific Discipline Formation*, ed. W. R. Woodward and R. S. Cohen (Amsterdam: Kluwer Academic, 1991), pp. 141–150.]

p. 47, l. 9 [This super-acute microscopic observer was first considered by Maxwell and is often called "Maxwell's demon." See Brush, *Heat*, vol. 2, pp. 589–593 and Martin J. Klein, "Maxwell, His Demon, and the Second Law of Thermodynamics," *American Scientist* **58**:1, 84–97 (1970).]

p. 50, l. 7b ff [In this concept of "elementary disorder" Planck shows his fundamental departure from Boltzmann's statistical approach. For Boltzmann nothing could prevent the *possibility* of the decrease of entropy, however unlikely; for Planck, the presumption of elementary disorder makes any such decrease impossible. This disorder mandates a new kind of counting of states, for it rests on the radical identicality which Planck introduces on pp. 52–53 and uses on pp. 90–91. On the differences with Boltzmann see Needell, "Introduction," pp. xix–xxi and "Irreversibility," pp. 42–55.]

p. 51, n. 1 [Here Planck refers to Henri Poincaré's proof (1890) that a finite system of bodies governed by classical mechanics will eventually reach every possible state an infinite number of times. Such "eternal return" implies that the entropy must periodically return to a given starting point, violating the second law, which calls for constant increase of entropy. This theorem had been at the center of a dispute between Ernst Zermelo, a student of Planck's, and Ludwig Boltzmann (1896). Boltzmann's resolution was that entropy does not always increase, only almost always; Planck here also notes that Poincaré had assumed ideally smooth walls bounding the system. See Brush, *Heat*, vol. 1, pp. 238–240, 627; for the relevant original papers of Poincaré, Zermelo, and Boltzmann see Brush, *Kinetic Theory*, vol. 2, pp. 194–245.]

p. 52, l. 9b [Planck here makes explicit his awareness of the fundamental significance of the sameness of these elements, which James Clerk Maxwell also had recognized in his *Theory of Heat* (New York: D. Appleton, 1872), pp. 309–312. See Pesic, "Identicality," pp. 975–977.]

p. 55, l. 6b [Implicitly entropy is an "extensive" quantity, proportional to the amount of substance. It is contrasted with an "intensive" quantity like temperature which does not depend on the quantity of substance.]

p. 55, eq. (12) [Planck introduced k to specify the constant of proportionality and named it after Boltzmann. Note that if k were not constant, entropy would no longer be simply additive for independent states.]

p. 56, l. 13 [Gibbs called this manifold of state domains "extension-in-phase" but it is now usually called "phase space"; its coordinates are the generalized coordinates and momenta for all the particles. If there are N particles, each of which has 3 coordinates and 3 momenta, then there are $6N$ dimensions in phase space. The probability of a given state of such a system can now be defined in a geometric way as the volume that the state occupies in phase space (see the last equation on p. 57). Note that this volume is tacitly compared to the whole physically possible volume in phase space the system might ever occupy (defined to have volume = total probability = 1). As Planck mentions, this volume is constant in time (Liouville's Theorem); see Brush, *Heat*, vol. 2, p. 634.]

FOURTH LECTURE

p. 60, l. 3 [See the note to p. 44. Planck, following Boltzmann, lets the size of the elementary domains go to zero at the end of the calculation, leading to the Maxwellian distribution (p. 63); in the case of black-body radiation Planck does not take this limit and thus arrives at his new distribution (p. 92).]

p. 68, l. 6 [This "uniform distribution" is now usually called "equipartition of energy" when applied in general to any degree of freedom of the system. It does not hold in Planck's treatment of radiation, as he notes on p. 92, l. 4. Planck's suspicions about the experimental values of specific heats were quickly borne out; see Stephen G. Brush, *Statistical Physics and the Atomic Theory of Matter* (Princeton: Princeton University Press, 1983), pp. 145–148.]

FIFTH LECTURE

p. 74, l. 18 [*Diathermanous* means allowing radiant energy to pass without absorption; here, the vacuum allows all frequencies of light to pass.]

p. 75, n. 1 [See Planck, *Theory of Heat Radiation*, pp. 1–45.]

p. 81, l. 6b [Here Planck introduces his idealized oscillator, which is not explicitly a realistic model of an atom, but rather the simplest way of representing the universal qualities of radiation in a black body. Using this oscillator (which he will call a "resonator" in the next lecture) Planck envisages a simple thought-experiment to probe the radiation.]

p. 83, l. 3b [This equation represents the radiation such an oscillator must emit solely as a consequence of Maxwell's equations. Planck does not introduce any internal damping of the oscillator, for that would require introducing friction *ad hoc*. His original hope had been that this radiation loss might exactly account for the increasing entropy of the radiation.]

p. 86, l. 4 [By "spectral division" Planck means rewriting the preceding equation to show the contribution to the sum over Fourier coefficients near the oscillator's natural frequency ν_o. (*PAV* 1.758–762); for a helpful summary of this argument see Jammer, *Conceptual Foundations*, Appendix A, pp. 383–385 and Kangro, *Early History*, pp. 138–141.]

SIXTH LECTURE

p. 87, ll. 5–7 [Planck now imagines putting an oscillator into the cavity of the black body to probe the intensity of the radiation; he calls it a "resonator" to emphasize its resonant response to the ambient radiation. He is seeking the relation between the intensity of radiation K_ν and the temperature T and frequency ν of that radiation. This function K_ν governs the "normal spectrum." In order to find it, his strategy is to find the entropy of that probe-resonator in terms of the energy of the surrounding radiation U.]

p. 87, eq. (48) [Planck's derivation of eq. (48) in the Fifth Lecture can be rendered more intuitive by a dimensional argument. The dimensions of K_ν, the specific intensity of radiation for frequency ν, are energy per cm^2. If the resonator has energy U and is in equilibrium with the ambient radiation the simplest expression having the right dimensions is $K_\nu \propto U/\lambda^2 = (\nu^2/c^2)U$, since the available length is the wavelength $\lambda = c/\nu$, where c is the speed of light. In the Fifth Lecture, a detailed analysis that takes into account the electrodynamic effects experienced by an oscillator immersed in a radiation field gives exactly this expression.]

p. 87, eq. (49) [The entropy is $dS = dQ/T$ and in the cavity $dQ = dU$.]

p. 88, l. 12 [Planck will calculate the probability associated with the resonator having energy U as a result of being bathed in the radiation inside the cavity. Then he will calculate the entropy associated with this probability (as a function of U and ν), and this he will put into eq. (49) to find the energy U (as a function of ν and T). This in turn he will put into eq. (48) to determine K_ν.]

p. 88, l. 2b-1b [He calls the generalized momentum the "impulse" here.]

p. 89, l. 9 [These curves are ellipses because they have the form $x^2/a^2 + y^2/b^2 = 1$, where a and b are the semi-major and semi-minor axes. In general the area of such an ellipse is πab. Here, $a = \sqrt{2U/K}$ and $b = \sqrt{2UL}$ and the area of the ellipse is U/ν.]

p. 90, ll. 1-5 [The assumption that Planck makes is generally called the "ergodic hypothesis." Whether or in what sense it is true remains a question. In his form, it asserts that instead of averaging the behavior of one resonator over time, one can substitute for this the average at one time of many resonators spread over space. In another form, it asserts that a system will eventually pass through every state it can possibly assume consistent with its given total energy (or very close, if "quasi-ergodic"). For example, if a box of gas were really an ergodic system, eventually all of the molecules would pass through all the positions and speeds allowed by the constancy of the total energy of the gas. If this were true, the gas would eventually lose any "memory" of its exact initial state (in the course of the random visitation is makes of all possible states). Roughly speaking, what is at stake is the degree to which the system randomizes away its initial conditions (e.g., through the innumerable molecular collisions in a gas). For further discussion, see Brush, *Heat*, vol. 2, pp. 363–385.]

p. 90, ll. 8b-2b [Planck has stipulated that only integral multiples of the energy element ε can "fall upon" (i.e., be found in) a resonator. The N resonators must then share the P available energy elements. Note that Planck's way of writing this is to show by the numeral which resonator (i.e., mode of vibration of the cavity) is in question, and by the number of times with which it appears how many energy elements it has. Thus in his example, 1 1 3 3 3 4, resonator 1 appears twice and hence has two energy elements, number 2 does not appear (and has no energy elements), etc.]

p. 91, l. 9 [One might write Planck's example in this way: $\|\varepsilon\varepsilon|\;|\varepsilon\varepsilon\varepsilon|\varepsilon\|$, in which the divider symbol | separates the energy symbols ε for each of the different oscillators, numbered from left to right (note that oscillator number 2 has no energy elements and hence a blank | |). In Planck's example there are $N = 4$ oscillators and $P = 6$ energy elements ε present. Note also that there will always be $(N-1)$ divider symbols | when there are N oscillators. Now there are $(P+N-1)$ symbols ε and | to be found in every complexion pertaining to the given state. Since the number of simple permutations of n things is given by $n! = 1 \times 2 \times 3 \times \ldots \times (n-1) \times n$ the $(P+N-1)$ symbols for any state can be found in $(P+N-1)!$ simple permutations. But any of the $P!$ permutations of the P energy symbols ε among themselves do not change the complexion of the state, nor do the $(N-1)!$ permutations of the $(N-1)$ dividers among themselves (i.e., just interchanging the *dividers*, not the elements they divide). Thus these factors must be divided out of the total number of simple permutations $(P+N-1)!$, yielding Planck's result, $W = (P+N-1)!/(N-1)!P!$. This witty proof, with its inspired use of the divider | to render transparent the counting of the different cases, is due to Paul Ehrenfest (1914); see his *Collected Scientific Papers*, ed. Martin J. Klein (Amsterdam: North-Holland, 1959), pp. 353–356 and Klein, *Ehrenfest*, pp. 255–257.]

p. 91, l. 5b [To obtain the "easy transformation", divide both sides of the expression for S_N by N. Then write the logarithms as $\log (N+P) = \log N(1+ P/N) = \log N + \log (1+ P/N)$ and likewise $\log P = \log N(P/N) = \log N + \log (P/N)$. Finally, note that $P/N = U/h\nu$. To gain a more intuitive understanding of the entropy of a single resonator, notice that if the frequency ν is very low the entropy becomes extremely large.]

p. 92, l. 4 [The classical prediction required that the total energy be equally distributed among all the possible modes of vibration (i.e., different frequencies) in the cavity (the so-called "equipartition of energy"). This can be seen as a limiting case of Planck's formula for U; to do this, use the Taylor series $e^x \cong 1+x$, which is good if $x \ll 1$, and take $x = h\nu/kT$, yielding $U \cong kT$ when $h\nu \ll kT$ (the limit of low-energy light radiation). Notice that the classical prediction predicts that the intensity should approach infinity as the frequency ν grows (as the light becomes more ultraviolet in color). It is clear from simple experiments on black-body radiation that the intensity does not rise indefinitely with rising frequency. Although Planck was not aware of this in 1909, shortly afterwards in 1911 Ehrenfest drew attention to what he called the "ultraviolet catastrophe" that had been implicit in classical electrodynamics; see Klein, *Ehrenfest*, pp. 249–250.]

p. 92, eq. (54) [Recall that $c = \lambda\nu$ and take $d\nu/d\lambda$.]

p. 93, l. 3 [In 1894 Wilhelm Wien had showed that the energy per unit volume of black-body radiation must have the form $\nu^3 G(\nu/T)$, where $G(\nu/T)$ is a function which can only depend on the ratio of the frequency ν to the temperature T in the cavity. He deduced this from Maxwell's equations and thermodynamics, although he could not give the form of this function G. His result implies that the distribution of energy is known for all temperatures when it has been determined for one temperature. Planck's result gives the exact form of the function G.]

p. 93, l. 7 [Lord Rayleigh's result dates from 1900 and followed from a consequence of classical statistical mechanics called the principle of equipartition (see note to p. 92, l. 4 above). He realized that it contradicted Wien's result and even speculated that the "doctrine" of equipartition might fail for "the graver modes" of vibration (i.e., those of longer wavelength). To show how Rayleigh's result emerges as an approximation to Planck's formula, use again the Taylor series $e^x \cong 1+x$ $(x \ll 1)$, and take $x = ch/k\lambda T$. Klein has argued that Planck only recognized Rayleigh's result in 1906; see "Planck, Entropy, and Quanta," pp. 99–101. Note also that in making this approximation Planck has only included the first term in the expansion $U \cong kT - h\nu/2$, showing that he did not yet recognize the "zero-point energy" $h\nu/2$ which he later included in his "second theory" of 1912; see P. W. Miloni and M.-L. Shih, "Zero-point energy in early quantum theory," *American Journal of Physics* **59**, 684–698 (1991) at 686–688.]

p. 93, l. 13 [Josef Stefan first established experimentally in 1879 that the energy density of black body radiation was aT^4, a result confirmed by the thermodynamic arguments of Boltzmann (1884). It is especially satisfying that Planck can calculate the Stefan-Boltzmann constant, a from the fundamental constants h, c, and k. The appearance of the infinite series α results from the evaluation of the integral in the total energy density, $\epsilon = (8\pi/q) \int_0^\infty K_\nu d\nu$; see F. Reif, *Fundamentals of statistical and thermal physics* (New York: McGraw-Hill, 1965), pp. 622–624.]

p. 94, l. 15 [By $1/\infty$ Planck denotes what is now called Avogadro's (or Loschmidt's) number, $N_0 = 6.023 \times 10^{23}$.]

p. 95, l. 15 [Jeans tried to eliminate h as a necessary component of Planck's theory. He argued that h was introduced as a stopgap in the calculation and that it should be allowed to go to zero. If this is done, the classical result is obtained but the total energy in the cavity is no longer finite. By 1912 Ehrenfest and Poincaré had demonstrated that the quantization $E = h\nu$ is a *necessary* as well as sufficient condition for the finiteness of the total energy of the cavity. At that point Jeans conceded the fundamental significance of h, which Planck had sensed from the beginning; see Geoffrey Gorham, "Planck's Principle and Jeans's Conversion," *Studies in History and Philosophy of Science*, **22**, 471–497 (1991).]

p. 95, l. 6b [Here Planck responds to a paper Einstein published on March 15, 1909 "On the Present Status of the Radiation Problem," in *The Collected Papers of Albert Einstein*, ed. John Stachel (Princeton: Princeton University Press, 1989), vol. 2, pp. 541–553 [English translation: pp. 357–375], at 544–548 [361–371]; Planck elaborated his reaction in a discussion on November 10, 1909 (pp. 584–587 [395–398]), upon his return to Germany. See Needell, "Irreversibility," pp. 113–131.]

SEVENTH LECTURE

p. 99, l. 13 [In his work on relativity in 1906 Planck relied on the principle of least action to guide his considerations and was the first to cast Einstein's work into this form (*PAV* 2.176–209). Thus although this lecture is not limited to the relativistic case, Planck's considerations here are extremely closely linked to those of the succeeding lecture on relativity. See Goldberg, "Planck's Philosophy," pp. 130–131.]

p. 107, l. 4 [But compare p. 97, l. 11b: "in the final analysis all processes are reversible." Planck's hesitation about the depth of irreversibility shows the complexity of the question in his mind.]

EIGHTH LECTURE

p. 112, l. 6b [There was practically no response to relativity in America before 1908; see Goldberg, *Understanding Relativity*

(Boston: Birkhäuser, 1984), pp. 248–251. The year following these lectures Planck published a brief popular account of relativity for the American audience, "The Mechanical Theory of Nature," *Scientific American Supplement* **70**, 387 (December 17, 1910).]

p. 119, l. 18 [Although Planck was the first major physicist to advocate Einstein's theory, he hesitated to remove the ether from physical theory. During 1906–1908 he tried to apply relativity only to the mechanical level of the theory without completely abolishing the ether (*PAV* 2.115–120). However, here he gives his first full account of his new understanding of the significance of the invariance of the speed of light, as noted by Goldberg, "Planck's Philosophy," pp. 157–158, n. 116.]

p. 123, l. 4b [Planck here is one of the first to draw attention to the four-dimensional space-time which Minkowski eloquently presented in his 1908 lecture "Space and Time," reprinted in *The Principle of Relativity*, pp. 75–91.]

p. 128, l. 4b [For Hasenöhrl's work and Planck's response, see Goldberg, "Planck's Philosophy," pp. 133–142.]

p. 129, l. 12b [In section 8 of his 1905 electrodynamics paper Einstein had noted that "it is remarkable that the energy and the frequency of a light complex vary with the state of motion of the observer in accordance with the same law" (*The Principle of Relativity*, p. 58). Though this implies the invariance of E/ν, Planck's 1907 demonstration of the relativistic invariance of action was the first explicit application of relativity to the quantum theory (*PAV* 2.176–209). As a consequence, the constant h is relativistically invariant. Planck noted that "because of this theorem the significance of the principle of least action is extended in a new direction" (*PAV* 2.198); see Abraham Pais, *Subtle is the Lord . . .* (Oxford: Oxford University Press, 1982), p. 151. Indeed, variational principles were invoked at several stages of the "old quantum theory" and were a crucial element in Schrödinger's original derivation of the wave equation; see Yourgrau and Mandelstam, *Variational Principles*, pp. 97–126 and Erwin Schrödinger, *Four Lectures on Wave Mechanics* (London: Blackie & Son, 1928), pp. 1–10.]

A CATALOG OF SELECTED

DOVER BOOKS

IN ALL FIELDS OF INTEREST

A CATALOG OF SELECTED DOVER BOOKS IN ALL FIELDS OF INTEREST

CONCERNING THE SPIRITUAL IN ART, Wassily Kandinsky. Pioneering work by father of abstract art. Thoughts on color theory, nature of art. Analysis of earlier masters. 12 illustrations. 80pp. of text. 5⅜ x 8½. 23411-8 Pa. $3.95

ANIMALS: 1,419 Copyright-Free Illustrations of Mammals, Birds, Fish, Insects, etc., Jim Harter (ed.). Clear wood engravings present, in extremely lifelike poses, over 1,000 species of animals. One of the most extensive pictorial sourcebooks of its kind. Captions. Index. 284pp. 9 x 12. 23766-4 Pa. $12.95

CELTIC ART: The Methods of Construction, George Bain. Simple geometric techniques for making Celtic interlacements, spirals, Kells-type initials, animals, humans, etc. Over 500 illustrations. 160pp. 9 x 12. (USO) 22923-8 Pa. $9.95

AN ATLAS OF ANATOMY FOR ARTISTS, Fritz Schider. Most thorough reference work on art anatomy in the world. Hundreds of illustrations, including selections from works by Vesalius, Leonardo, Goya, Ingres, Michelangelo, others. 593 illustrations. 192pp. 7⅛ x 10¼. 20241-0 Pa. $9.95

CELTIC HAND STROKE-BY-STROKE (Irish Half-Uncial from "The Book of Kells"): An Arthur Baker Calligraphy Manual, Arthur Baker. Complete guide to creating each letter of the alphabet in distinctive Celtic manner. Covers hand position, strokes, pens, inks, paper, more. Illustrated. 48pp. 8¼ x 11. 24336-2 Pa. $3.95

EASY ORIGAMI, John Montroll. Charming collection of 32 projects (hat, cup, pelican, piano, swan, many more) specially designed for the novice origami hobbyist. Clearly illustrated easy-to-follow instructions insure that even beginning papercrafters will achieve successful results. 48pp. 8¼ x 11. 27298-2 Pa. $3.50

THE COMPLETE BOOK OF BIRDHOUSE CONSTRUCTION FOR WOODWORKERS, Scott D. Campbell. Detailed instructions, illustrations, tables. Also data on bird habitat and instinct patterns. Bibliography. 3 tables. 63 illustrations in 15 figures. 48pp. 5¼ x 8½. 24407-5 Pa. $2.50

BLOOMINGDALE'S ILLUSTRATED 1886 CATALOG: Fashions, Dry Goods and Housewares, Bloomingdale Brothers. Famed merchants' extremely rare catalog depicting about 1,700 products: clothing, housewares, firearms, dry goods, jewelry, more. Invaluable for dating, identifying vintage items. Also, copyright-free graphics for artists, designers. Co-published with Henry Ford Museum & Greenfield Village. 160pp. 8¼ x 11. 25780-0 Pa. $10.95

HISTORIC COSTUME IN PICTURES, Braun & Schneider. Over 1,450 costumed figures in clearly detailed engravings–from dawn of civilization to end of 19th century. Captions. Many folk costumes. 256pp. 8⅝ x 11¾. 23150-X Pa. $12.95

BRASS INSTRUMENTS: Their History and Development, Anthony Baines. Authoritative, updated survey of the evolution of trumpets, trombones, bugles, cornets, French horns, tubas and other brass wind instruments. Over 140 illustrations and 48 music examples. Corrected and updated by author. New preface. Bibliography. 320pp. 5⅜ x 8½. 27574-4 Pa. $9.95

HOLLYWOOD GLAMOR PORTRAITS, John Kobal (ed.). 145 photos from 1926-49. Harlow, Gable, Bogart, Bacall; 94 stars in all. Full background on photographers, technical aspects. 160pp. 8⅜ x 11¼. 23352-9 Pa. $12.95

MAX AND MORITZ, Wilhelm Busch. Great humor classic in both German and English. Also 10 other works: "Cat and Mouse," "Plisch and Plumm," etc. 216pp. 5⅜ x 8½. 20181-3 Pa. $6.95

THE RAVEN AND OTHER FAVORITE POEMS, Edgar Allan Poe. Over 40 of the author's most memorable poems: "The Bells," "Ulalume," "Israfel," "To Helen," "The Conqueror Worm," "Eldorado," "Annabel Lee," many more. Alphabetic lists of titles and first lines. 64pp. 5$\frac{3}{16}$ x 8¼. 26685-0 Pa. $1.00

PERSONAL MEMOIRS OF U. S. GRANT, Ulysses Simpson Grant. Intelligent, deeply moving firsthand account of Civil War campaigns, considered by many the finest military memoirs ever written. Includes letters, historic photographs, maps and more. 528pp. 6⅛ x 9¼. 28587-1 Pa. $11.95

AMULETS AND SUPERSTITIONS, E. A. Wallis Budge. Comprehensive discourse on origin, powers of amulets in many ancient cultures: Arab, Persian Babylonian, Assyrian, Egyptian, Gnostic, Hebrew, Phoenician, Syriac, etc. Covers cross, swastika, crucifix, seals, rings, stones, etc. 584pp. 5⅜ x 8½. 23573-4 Pa. $12.95

RUSSIAN STORIES/PYCCKNE PACCKA3bl: A Dual-Language Book, edited by Gleb Struve. Twelve tales by such masters as Chekhov, Tolstoy, Dostoevsky, Pushkin, others. Excellent word-for-word English translations on facing pages, plus teaching and study aids, Russian/English vocabulary, biographical/critical introductions, more. 416pp. 5⅜ x 8½. 26244-8 Pa. $8.95

PHILADELPHIA THEN AND NOW: 60 Sites Photographed in the Past and Present, Kenneth Finkel and Susan Oyama. Rare photographs of City Hall, Logan Square, Independence Hall, Betsy Ross House, other landmarks juxtaposed with contemporary views. Captures changing face of historic city. Introduction. Captions. 128pp. 8¼ x 11. 25790-8 Pa. $9.95

AIA ARCHITECTURAL GUIDE TO NASSAU AND SUFFOLK COUNTIES, LONG ISLAND, The American Institute of Architects, Long Island Chapter, and the Society for the Preservation of Long Island Antiquities. Comprehensive, well-researched and generously illustrated volume brings to life over three centuries of Long Island's great architectural heritage. More than 240 photographs with authoritative, extensively detailed captions. 176pp. 8¼ x 11. 26946-9 Pa. $14.95

NORTH AMERICAN INDIAN LIFE: Customs and Traditions of 23 Tribes, Elsie Clews Parsons (ed.). 27 fictionalized essays by noted anthropologists examine religion, customs, government, additional facets of life among the Winnebago, Crow, Zuni, Eskimo, other tribes. 480pp. 6⅛ x 9¼. 27377-6 Pa. $10.95

THE BEST TALES OF HOFFMANN, E. T. A. Hoffmann. 10 of Hoffmann's most important stories: "Nutcracker and the King of Mice," "The Golden Flowerpot," etc. 458pp. 5⅜ x 8½. 21793-0 Pa. $9.95

FROM FETISH TO GOD IN ANCIENT EGYPT, E. A. Wallis Budge. Rich detailed survey of Egyptian conception of "God" and gods, magic, cult of animals, Osiris, more. Also, superb English translations of hymns and legends. 240 illustrations. 545pp. 5⅜ x 8½. 25803-3 Pa. $13.95

FRENCH STORIES/CONTES FRANÇAIS: A Dual-Language Book, Wallace Fowlie. Ten stories by French masters, Voltaire to Camus: "Micromegas" by Voltaire; "The Atheist's Mass" by Balzac; "Minuet" by de Maupassant; "The Guest" by Camus, six more. Excellent English translations on facing pages. Also French-English vocabulary list, exercises, more. 352pp. 5⅜ x 8½. 26443-2 Pa. $8.95

CHICAGO AT THE TURN OF THE CENTURY IN PHOTOGRAPHS: 122 Historic Views from the Collections of the Chicago Historical Society, Larry A. Viskochil. Rare large-format prints offer detailed views of City Hall, State Street, the Loop, Hull House, Union Station, many other landmarks, circa 1904-1913. Introduction. Captions. Maps. 144pp. 9⅜ x 12¼. 24656-6 Pa. $12.95

OLD BROOKLYN IN EARLY PHOTOGRAPHS, 1865-1929, William Lee Younger. Luna Park, Gravesend race track, construction of Grand Army Plaza, moving of Hotel Brighton, etc. 157 previously unpublished photographs. 165pp. 8⅞ x 11¾. 23587-4 Pa. $13.95

THE MYTHS OF THE NORTH AMERICAN INDIANS, Lewis Spence. Rich anthology of the myths and legends of the Algonquins, Iroquois, Pawnees and Sioux, prefaced by an extensive historical and ethnological commentary. 36 illustrations. 480pp. 5⅜ x 8½. 25967-6 Pa. $8.95

AN ENCYCLOPEDIA OF BATTLES: Accounts of Over 1,560 Battles from 1479 B.C. to the Present, David Eggenberger. Essential details of every major battle in recorded history from the first battle of Megiddo in 1479 B.C. to Grenada in 1984. List of Battle Maps. New Appendix covering the years 1967-1984. Index. 99 illustrations. 544pp. 6½ x 9¼. 24913-1 Pa. $14.95

SAILING ALONE AROUND THE WORLD, Captain Joshua Slocum. First man to sail around the world, alone, in small boat. One of great feats of seamanship told in delightful manner. 67 illustrations. 294pp. 5⅜ x 8½. 20326-3 Pa. $5.95

ANARCHISM AND OTHER ESSAYS, Emma Goldman. Powerful, penetrating, prophetic essays on direct action, role of minorities, prison reform, puritan hypocrisy, violence, etc. 271pp. 5⅜ x 8½. 22484-8 Pa. $6.95

MYTHS OF THE HINDUS AND BUDDHISTS, Ananda K. Coomaraswamy and Sister Nivedita. Great stories of the epics; deeds of Krishna, Shiva, taken from puranas, Vedas, folk tales; etc. 32 illustrations. 400pp. 5⅜ x 8½. 21759-0 Pa. $10.95

BEYOND PSYCHOLOGY, Otto Rank. Fear of death, desire of immortality, nature of sexuality, social organization, creativity, according to Rankian system. 291pp. 5⅜ x 8½. 20485-5 Pa. $8.95

A THEOLOGICO-POLITICAL TREATISE, Benedict Spinoza. Also contains unfinished Political Treatise. Great classic on religious liberty, theory of government on common consent. R. Elwes translation. Total of 421pp. 5⅜ x 8½. 20249-6 Pa. $9.95

PERSPECTIVE FOR ARTISTS, Rex Vicat Cole. Depth, perspective of sky and sea, shadows, much more, not usually covered. 391 diagrams, 81 reproductions of drawings and paintings. 279pp. 5⅜ x 8½. 22487-2 Pa. $7.95

DRAWING THE LIVING FIGURE, Joseph Sheppard. Innovative approach to artistic anatomy focuses on specifics of surface anatomy, rather than muscles and bones. Over 170 drawings of live models in front, back and side views, and in widely varying poses. Accompanying diagrams. 177 illustrations. Introduction. Index. 144pp. 8⅜ x11¼. 26723-7 Pa. $8.95

GOTHIC AND OLD ENGLISH ALPHABETS: 100 Complete Fonts, Dan X. Solo. Add power, elegance to posters, signs, other graphics with 100 stunning copyright-free alphabets: Blackstone, Dolbey, Germania, 97 more–including many lower-case, numerals, punctuation marks. 104pp. 8⅛ x 11. 24695-7 Pa. $8.95

HOW TO DO BEADWORK, Mary White. Fundamental book on craft from simple projects to five-bead chains and woven works. 106 illustrations. 142pp. 5⅜ x 8. 20697-1 Pa. $4.95

THE BOOK OF WOOD CARVING, Charles Marshall Sayers. Finest book for beginners discusses fundamentals and offers 34 designs. "Absolutely first rate . . . well thought out and well executed."–E. J. Tangerman. 118pp. 7¾ x 10⅝. 23654-4 Pa. $6.95

ILLUSTRATED CATALOG OF CIVIL WAR MILITARY GOODS: Union Army Weapons, Insignia, Uniform Accessories, and Other Equipment, Schuyler, Hartley, and Graham. Rare, profusely illustrated 1846 catalog includes Union Army uniform and dress regulations, arms and ammunition, coats, insignia, flags, swords, rifles, etc. 226 illustrations. 160pp. 9 x 12. 24939-5 Pa. $10.95

WOMEN'S FASHIONS OF THE EARLY 1900s: An Unabridged Republication of "New York Fashions, 1909," National Cloak & Suit Co. Rare catalog of mail-order fashions documents women's and children's clothing styles shortly after the turn of the century. Captions offer full descriptions, prices. Invaluable resource for fashion, costume historians. Approximately 725 illustrations. 128pp. 8⅜ x 11¼. 27276-1 Pa. $11.95

THE 1912 AND 1915 GUSTAV STICKLEY FURNITURE CATALOGS, Gustav Stickley. With over 200 detailed illustrations and descriptions, these two catalogs are essential reading and reference materials and identification guides for Stickley furniture. Captions cite materials, dimensions and prices. 112pp. 6½ x 9¼. 26676-1 Pa. $9.95

EARLY AMERICAN LOCOMOTIVES, John H. White, Jr. Finest locomotive engravings from early 19th century: historical (1804–74), main-line (after 1870), special, foreign, etc. 147 plates. 142pp. 11⅜ x 8¼. 22772-3 Pa. $10.95

THE TALL SHIPS OF TODAY IN PHOTOGRAPHS, Frank O. Braynard. Lavishly illustrated tribute to nearly 100 majestic contemporary sailing vessels: Amerigo Vespucci, Clearwater, Constitution, Eagle, Mayflower, Sea Cloud, Victory, many more. Authoritative captions provide statistics, background on each ship. 190 black-and-white photographs and illustrations. Introduction. 128pp. 8⅞ x 11¾. 27163-3 Pa. $13.95

PHOTOGRAPHIC SKETCHBOOK OF THE CIVIL WAR, Alexander Gardner. 100 photos taken on field during the Civil War. Famous shots of Manassas Harper's Ferry, Lincoln, Richmond, slave pens, etc. 244pp. 10⅝ x 8¼. 22731-6 Pa. $9.95

FIVE ACRES AND INDEPENDENCE, Maurice G. Kains. Great back-to-the-land classic explains basics of self-sufficient farming. The one book to get. 95 illustrations. 397pp. 5⅜ x 8½. 20974-1 Pa. $7.95

SONGS OF EASTERN BIRDS, Dr. Donald J. Borror. Songs and calls of 60 species most common to eastern U.S.: warblers, woodpeckers, flycatchers, thrushes, larks, many more in high-quality recording. Cassette and manual 99912-2 $9.95

A MODERN HERBAL, Margaret Grieve. Much the fullest, most exact, most useful compilation of herbal material. Gigantic alphabetical encyclopedia, from aconite to zedoary, gives botanical information, medical properties, folklore, economic uses, much else. Indispensable to serious reader. 161 illustrations. 888pp. 6½ x 9¼. 2-vol. set. (USO) Vol. I: 22798-7 Pa. $9.95
Vol. II: 22799-5 Pa. $9.95

HIDDEN TREASURE MAZE BOOK, Dave Phillips. Solve 34 challenging mazes accompanied by heroic tales of adventure. Evil dragons, people-eating plants, blood-thirsty giants, many more dangerous adversaries lurk at every twist and turn. 34 mazes, stories, solutions. 48pp. 8¼ x 11. 24566-7 Pa. $2.95

LETTERS OF W. A. MOZART, Wolfgang A. Mozart. Remarkable letters show bawdy wit, humor, imagination, musical insights, contemporary musical world; includes some letters from Leopold Mozart. 276pp. 5⅜ x 8½. 22859-2 Pa. $7.95

BASIC PRINCIPLES OF CLASSICAL BALLET, Agrippina Vaganova. Great Russian theoretician, teacher explains methods for teaching classical ballet. 118 illustrations. 175pp. 5⅜ x 8½. 22036-2 Pa. $5.95

THE JUMPING FROG, Mark Twain. Revenge edition. The original story of The Celebrated Jumping Frog of Calaveras County, a hapless French translation, and Twain's hilarious "retranslation" from the French. 12 illustrations. 66pp. 5⅜ x 8½. 22686-7 Pa. $3.95

BEST REMEMBERED POEMS, Martin Gardner (ed.). The 126 poems in this superb collection of 19th- and 20th-century British and American verse range from Shelley's "To a Skylark" to the impassioned "Renascence" of Edna St. Vincent Millay and to Edward Lear's whimsical "The Owl and the Pussycat." 224pp. 5⅜ x 8½. 27165-X Pa. $4.95

COMPLETE SONNETS, William Shakespeare. Over 150 exquisite poems deal with love, friendship, the tyranny of time, beauty's evanescence, death and other themes in language of remarkable power, precision and beauty. Glossary of archaic terms. 80pp. 5³⁄₁₆ x 8¼. 26686-9 Pa. $1.00

BODIES IN A BOOKSHOP, R. T. Campbell. Challenging mystery of blackmail and murder with ingenious plot and superbly drawn characters. In the best tradition of British suspense fiction. 192pp. 5⅜ x 8½. 24720-1 Pa. $6.95

THE WIT AND HUMOR OF OSCAR WILDE, Alvin Redman (ed.). More than 1,000 ripostes, paradoxes, wisecracks: Work is the curse of the drinking classes; I can resist everything except temptation; etc. 258pp. 5⅜ x 8½. 20602-5 Pa. $5.95

SHAKESPEARE LEXICON AND QUOTATION DICTIONARY, Alexander Schmidt. Full definitions, locations, shades of meaning in every word in plays and poems. More than 50,000 exact quotations. 1,485pp. 6½ x 9¼. 2-vol. set.
Vol. 1: 22726-X Pa. $16.95
Vol. 2: 22727-8 Pa. $16.95

SELECTED POEMS, Emily Dickinson. Over 100 best-known, best-loved poems by one of America's foremost poets, reprinted from authoritative early editions. No comparable edition at this price. Index of first lines. 64pp. 5³⁄₁₆ x 8¼.
26466-1 Pa. $1.00

CELEBRATED CASES OF JUDGE DEE (DEE GOONG AN), translated by Robert van Gulik. Authentic 18th-century Chinese detective novel; Dee and associates solve three interlocked cases. Led to van Gulik's own stories with same characters. Extensive introduction. 9 illustrations. 237pp. 5⅜ x 8½. 23337-5 Pa. $6.95

THE MALLEUS MALEFICARUM OF KRAMER AND SPRENGER, translated by Montague Summers. Full text of most important witchhunter's "bible," used by both Catholics and Protestants. 278pp. 6⅝ x 10. 22802-9 Pa. $12.95

SPANISH STORIES/CUENTOS ESPAÑOLES: A Dual-Language Book, Angel Flores (ed.). Unique format offers 13 great stories in Spanish by Cervantes, Borges, others. Faithful English translations on facing pages. 352pp. 5⅜ x 8½.
25399-6 Pa. $8.95

THE CHICAGO WORLD'S FAIR OF 1893: A Photographic Record, Stanley Appelbaum (ed.). 128 rare photos show 200 buildings, Beaux-Arts architecture, Midway, original Ferris Wheel, Edison's kinetoscope, more. Architectural emphasis; full text. 116pp. 8¼ x 11. 23990-X Pa. $9.95

OLD QUEENS, N.Y., IN EARLY PHOTOGRAPHS, Vincent F. Seyfried and William Asadorian. Over 160 rare photographs of Maspeth, Jamaica, Jackson Heights, and other areas. Vintage views of DeWitt Clinton mansion, 1939 World's Fair and more. Captions. 192pp. 8⅞ x 11. 26358-4 Pa. $12.95

CAPTURED BY THE INDIANS: 15 Firsthand Accounts, 1750-1870, Frederick Drimmer. Astounding true historical accounts of grisly torture, bloody conflicts, relentless pursuits, miraculous escapes and more, by people who lived to tell the tale. 384pp. 5⅜ x 8½. 24901-8 Pa. $8.95

THE WORLD'S GREAT SPEECHES, Lewis Copeland and Lawrence W. Lamm (eds.). Vast collection of 278 speeches of Greeks to 1970. Powerful and effective models; unique look at history. 842pp. 5⅜ x 8½. 20468-5 Pa. $14.95

THE BOOK OF THE SWORD, Sir Richard F. Burton. Great Victorian scholar/adventurer's eloquent, erudite history of the "queen of weapons"–from prehistory to early Roman Empire. Evolution and development of early swords, variations (sabre, broadsword, cutlass, scimitar, etc.), much more. 336pp. 6⅛ x 9¼.
25434-8 Pa. $9.95